MATHS PLUS 4

STAGE 2

MENTALS AND HOMEWORK BOOK

NEW SOUTH WALES SYLLABUS

Harry O'Brien
Greg Purcell

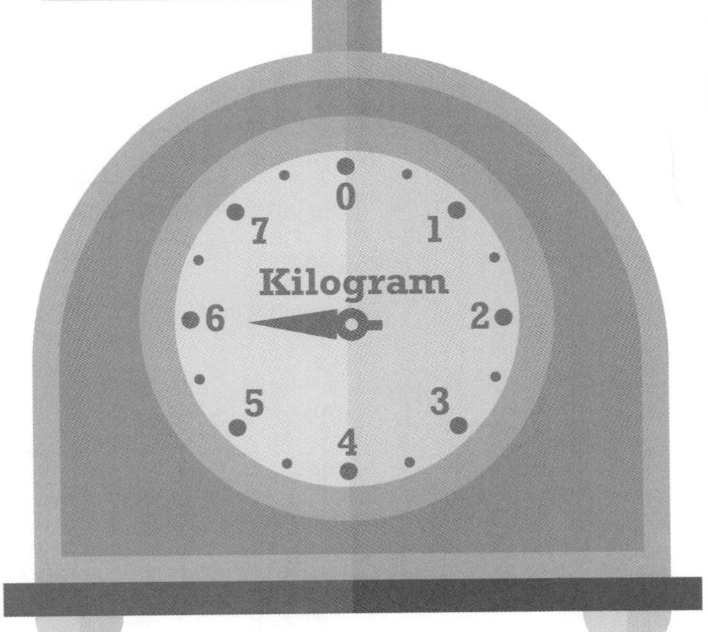

OXFORD
UNIVERSITY PRESS

Contents

Contents

NSW Syllabus Outcomes

Units	1	2	3	4	5	6	7
NUMBER AND ALGEBRA							
Representing numbers using place value							
MA2-RN-01 Applies an understanding of place value and the role of zero to represent numbers to at least tens of thousands							
MA2-RN-02 Represents and compares decimals up to 2 decimal places using place value							
Additive relations							
MA2-AR-01 Selects and uses mental and written strategies for addition and subtraction involving 2- and 3-digit numbers							
MA2-AR-02 Completes number sentences involving addition and subtraction by finding missing values							
Multiplicative relations							
MA2-MR-01 Represents and uses the structure of multiplicative relations to 10 x 10 to solve problems							
MA2-MR-02 Completes number sentences involving multiplication and division by finding missing values							
Partitioned fractions							
MA2-PF-01 Represents and compares halves, quarters, thirds and fifths as lengths on a number line and their related fractions formed by halving (eighths, sixths and tenths)							
MEASUREMENT AND SPACE							
Geometric measure							
MA2-GM-01 Uses grid maps and directional language to locate positions and follow routes							
MA2-GM-02 Measures and estimates lengths in metres, centimetres and millimetres							
MA2-GM-03 Identifies angles and classifies them by comparing to a right angle							
Two-dimensional spatial structure							
MA2-2DS-01 Compares two-dimensional shapes and describes their features							
MA2-2DS-02 Performs transformations by combining and splitting two-dimensional shapes							
MA2-2DS-03 Estimates, measures and compares areas using square centimetres and square metres							
Three-dimensional spatial structure							
MA2-3DS-01 Makes and sketches models and nets of three-dimensional objects including prisms and pyramids							
MA2-3DS-02 Estimates, measures and compares capacities (internal volumes) using litres, millilitres and volumes using cubic centimetres							
Non-spatial measure							
MA2-NSM-01 Estimates, measures and compares the masses of objects using kilograms and grams							
MA2-NSM-02 Represents and interprets analog and digital time in hours, minutes and seconds							
STATISTICS AND PROBABILITY							
Data							
MA2-DATA-01 Collects discrete data and constructs graphs using a given scale							
MA2-DATA-02 Interprets data in tables, dot plots and column graphs							
Chance							
MA2-CHAN-01 Records and compares the results of chance experiments							
MA0-WM-01 Working mathematically							
Develops understanding and fluency in mathematics through exploring and connecting mathematical concepts, choosing and applying mathematical techniques to solve problems, and communicating their thinking and reasoning coherently and clearly							

8	9	10	11	12	13	14	15	16	17	18	19	20	21	22	23	24	25	26	27	28	29	30	31	32	33	34	35
NUMBER AND ALGEBRA																											
Representing numbers using place value																											
Additive relations																											
Multiplicative relations																											
Partitioned fractions																											
MEASUREMENT AND SPACE																											
Geometric measure																											
Two-dimensional spatial structure																											
Three-dimensional spatial structure																											
Non-spatial measure																											
STATISTICS AND PROBABILITY																											
Data																											
Chance																											
MA0-WM-01 Working mathematically																											

UNIT 1

Number and Algebra

SET 1 Basic

1 4 + 2 + 6

2 6 + 8 + 3

3 9 − 5

4 20 − 5

5 6 × 5

6 7 × 4

7 2 × 10

8 Which is the third month of the year?

9 What is the sum of 9 and 5?

10 What is the difference between 11 and 8?

11 Subtract 5 from 11.

12 Halve 14c.

13 How many minutes in a quarter of an hour?

14

SET 2 Extending addition facts

Complete the facts.

1

+20	
30	
50	
10	
40	

2

+40	
40	
20	
30	
50	

3

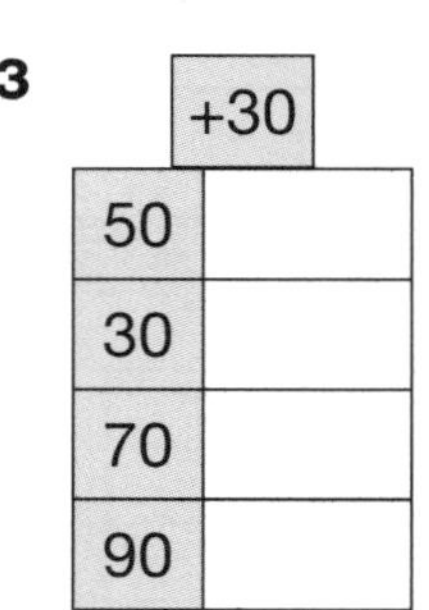

+30	
50	
30	
70	
90	

4

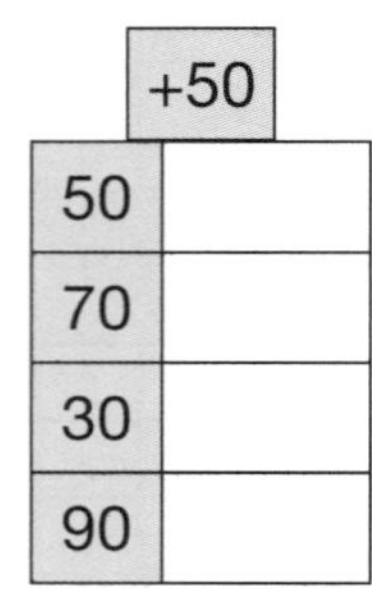

+50	
50	
70	
30	
90	

5 5 + 4

6 50 + 40

7 500 + 400

8 80 + 50

9 800 + 500

10 50 books plus 70 books

11 90 books plus 50 books

12 140 books plus 90 books

Space Three-dimensional revision

Colour code each matching description, name and shape.

1 My 2 bases are circles and my other face is curved.

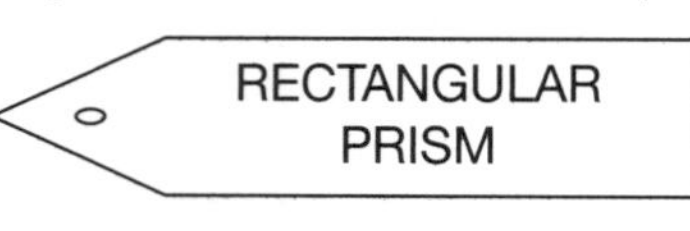

2 I have 6 faces which are all rectangles.

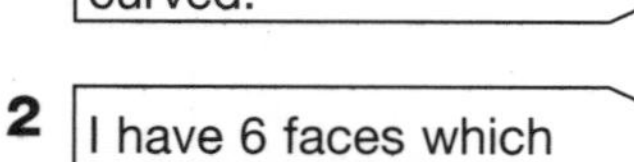

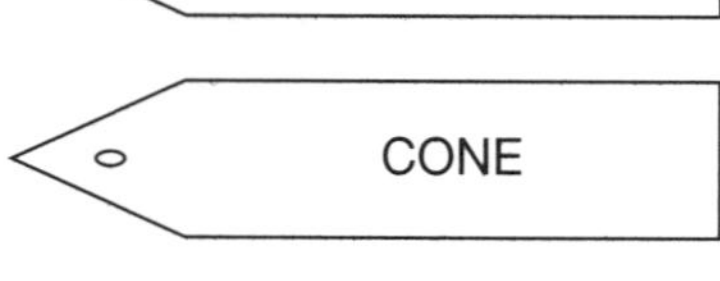

3 I have 1 base which is a circle.

CYLINDER

4 I have a square base and all my other faces are triangles.

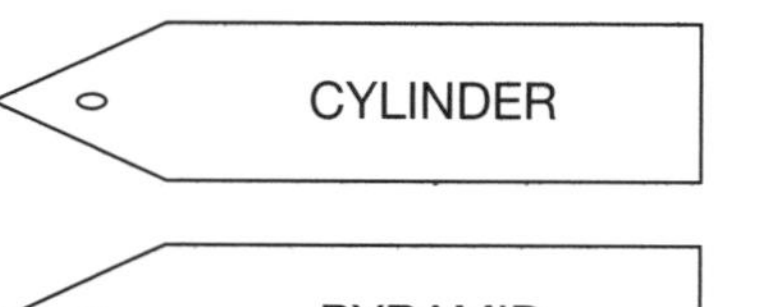

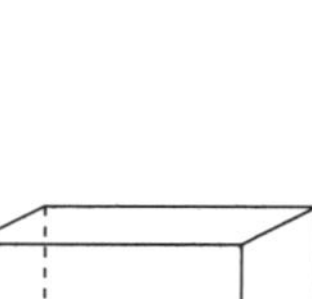

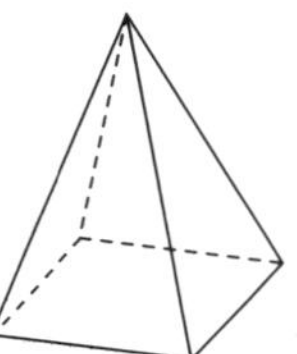

Number and Algebra

SET 3 Multiplication facts, times 3

1 3 books at $5 each = ☐

2 6 pens at $3 each = ☐

3 5 calculators at $3 each = ☐

4 7 rings at $3 each = ☐

5 8 bangles at $3 each = ☐

6 9 cups at $3 each = ☐

7 10 bags at $3 each = ☐

Complete the grids

8

× 5	
2	
7	
4	
6	
9	
8	

9

× 10	
2	
7	
4	
6	
9	
8	

10 Can you see any relationship between the times 5 and times 10 grids above? What is it?

____

____

SET 4 Extension

1 3 m and 17 cm = ☐ cm

2 274, 271, 268, ☐

3 How many centimetres in $4\frac{1}{2}$ m?

4 147, 151, 155, ☐

5 Value of 6 in 7631

6 $\frac{1}{4}$ of $72

7 What is 17 more than 191?

8 What is the sum of 237 and 122?

9 Halve 380.

10 How many sides on 3 octagons?

11 How much is $4\frac{1}{2}$ kg of meat at $12 per kilogram?

12 12 ÷ 3 + 37

13 How many tens in 376?

14 How many halves in 4 wholes?

15 Write 4207 in words.

____

____

16 List the factors of 48.

17 Write $\frac{3}{10}$ as a decimal.

18 Round 3866 to the nearest 100.

Measurement Centimetres

Extend these lines to the lengths listed.

5 cm

7 cm

9 cm

13 cm

$14\frac{1}{2}$ cm

$15\frac{1}{2}$ cm

Number and Algebra

SET 1 Basic

1 3 + ☐ = 10

2 8 + ☐ = 11

3 11 – 3

4 12 – ☐ = 8

5 2 × 10

6 3 × 2

7 4 × 6

8 ☐ × 5 = 20

9 What is the 2nd month of the year?

10 What is the product of 6 and 3?

11 What is the sum of 18 and 4?

12 What is the difference between 26 and 4?

13 How many minutes in 2 hours?

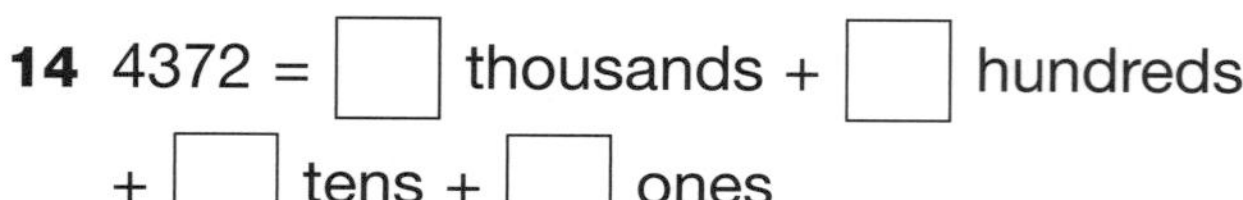

14 4372 = ☐ thousands + ☐ hundreds + ☐ tens + ☐ ones

15

SET 2 Subtraction

1

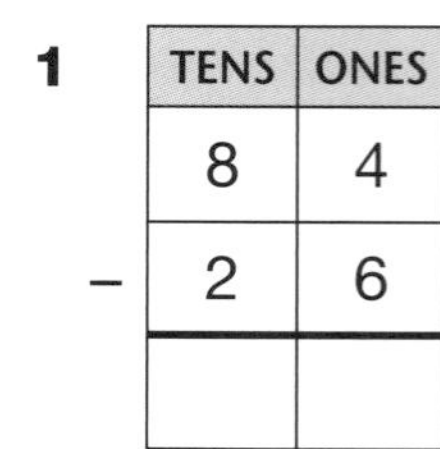

	TENS	ONES
	8	4
–	2	6

2

	TENS	ONES
	9	3
–	2	6

3

	TENS	ONES
	7	3
–	3	4

4

	TENS	ONES
	8	4
–	2	7

5

	TENS	ONES
	6	5
–	1	7

6

	TENS	ONES
	8	4
–	3	7

7

	HUND	TENS	ONES
	8	2	1
–	4	1	5

8

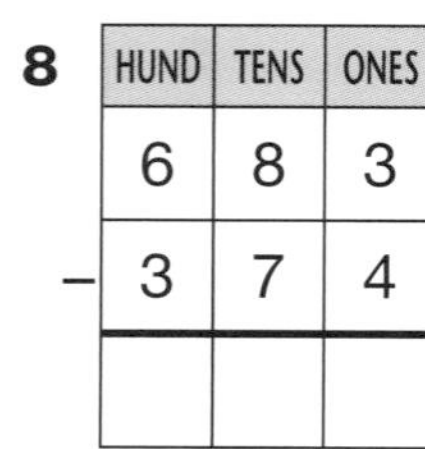

	HUND	TENS	ONES
	6	8	3
–	3	7	4

9

	HUND	TENS	ONES
	9	6	3
–	5	1	9

10 Hannah had $72 and spent $38. How much has she left?

11 There are 96 children in Ms Rosen's class. If 47 are boys, how many are girls?

Space Symmetry

Complete each shape using its line of symmetry.

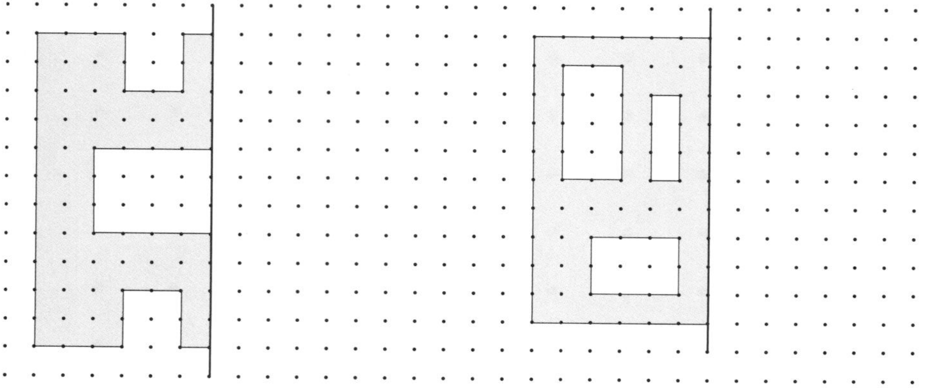

Number and Algebra

SET 3 Place value

State the place value of each bold number.

	Number	Place Value
1	**7**23	
2	8**4**2	
3	**3**746	
4	529**6**	
5	7**5**26	

6 What number is 100 more than 756?

7 What number is 100 more than 2357?

8 What number is 1000 more than 3576?

9 What number is 10 more than 2765?

10 What number is 100 less than 3574?

11 What number is 30 more than 367?

12 What number is 40 more than 1357?

13 What number is 50 less than 3590?

14 What number is 200 more than 3264?

15 1256 + 40

16 Write the largest number you can using 3, 7, 5.

17 Write the smallest number you can using 3, 7, 5.

SET 4 Extension

1 3 dozen take away 17

2 What is the difference between 42 and 8?

3 If today is Wednesday, what day is 11 days later?

4 50, 45, ☐, ☐, ☐, 25

5 6 × ☐ = 54

6 $5.63 = ☐ cents

7 What time is 4 hours after 3:45 pm?

8 What is left from $2.00 if I spent 85c?

9 How much are 8 pens at 90c each?

10 How much are 3 yo-yos at 25c each?

11 Make the largest number possible using 6, 4, 9, 8.

Working Mathematically

12 3 girls each paid $75 for their excursion. Name 3 sets of notes that equal $75.

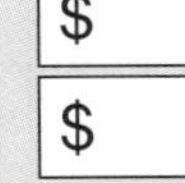

a	$	$	$	$	$
b	$	$	$	$	$
c	$	$	$	$	$

Measurement The square centimetre

Count the 1 cm² and $\frac{1}{2}$ cm² to work out the areas of the following shaded shapes.

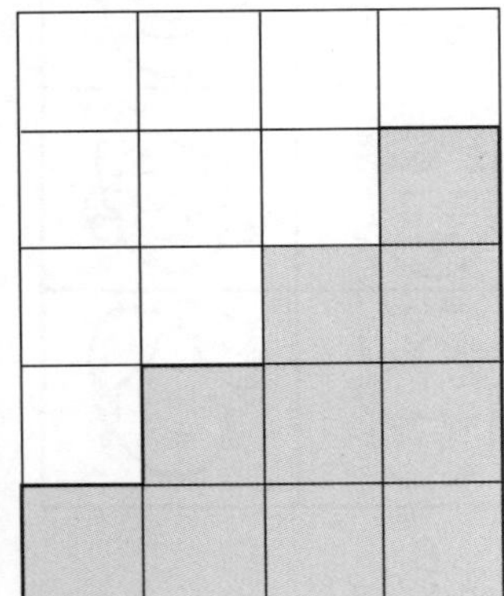

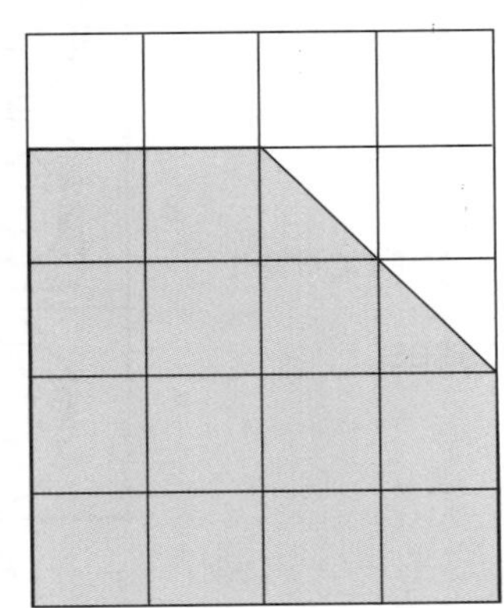

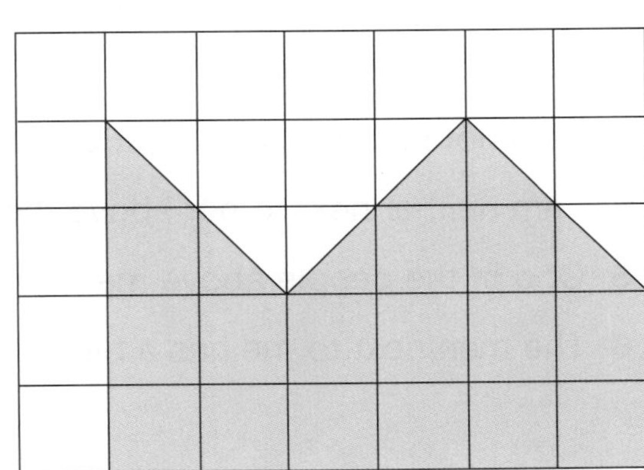

Number and Algebra

SET 1 Basic

1 5 + 6 + 8

2 9 − 7

3 4 × 2

4 (3 × 2) + 4

5 What is the product of 7 and 2?

6 What is the sum of 17 and 7?

7 4 + ☐ = 11

8 Double 9.

9 ☐ × 4 = 20

10 What is the product of 7 and 9?

11 3 + 3 + 3

12 Triple 5.

13 27 + 8

14 What is half of 36?

15

SET 2 2- and 3-digit addition

1

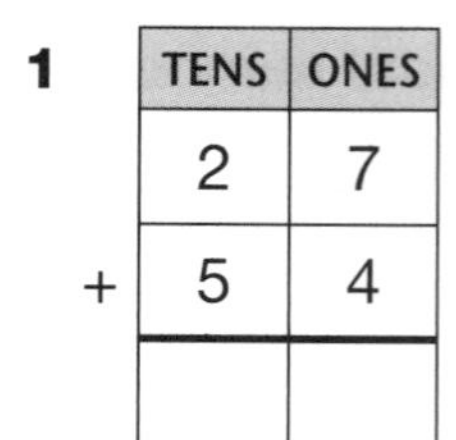

	TENS	ONES
	2	7
+	5	4

2

	TENS	ONES
	2	8
+	1	7

3

	TENS	ONES
	6	3
+	3	5

4

	TENS	ONES
	5	4
+	3	8

5

	TENS	ONES
	2	5
+	3	7

6

	TENS	ONES
	4	5
+	3	6

7

	HUND	TENS	ONES
	6	3	2
+	2	1	9

8

	HUND	TENS	ONES
	3	2	9
+	4	4	8

9

	TENS	ONES
	6	2
+	1	9

10 Use the answers above to find the missing word.

777	55	45	96	98
I	M	U	B	S

92	81	851	64	62
T	A	L	Y	R

__ __ __ __ __ __ __ __ __
1 2 3 4 5 6 7 8 9

Space Position

Which face is hiding the prize? Colour the correct face.

1 I do not have a moustache.

2 I am not in the middle.

3 I am not in the bottom right corner.

4 I am neither next to nor above the bottom right corner.

5 One of the people above me is wearing glasses.

6 The man next to me has a beard.

Number and Algebra

SET 3 Number patterns

Complete the number patterns.

1	5	10	15					
2	0	3	6					
3	20	30	40					
4	23	26	29					
5	216	218	220					
6	325	320	315					
7	500	450	400					
8	550	600	650					

Write the next number in the sequence.

9 245, 445, 645, ☐

10 1231, 2231, 3231, ☐

11 3107, 3307, 3507, ☐

SET 4 Extension

1 1 m and 27 cm = ☐ cm

2 What is the 6th letter of the alphabet?

3 How many $\frac{1}{4}$s in $1\frac{1}{2}$?

4 Share 36 among 4.

5 2 kg = ☐ g

6 Value of 2 in 3286

7 Multiply 3 by 2 and add 56.

8 3 thousands + 2 hundreds + 6 ones

9 2427 + 70

10 Round 1297 to the nearest 100.

11 27 more than 1351

12 If 3 kg cost $9, how much would 4 kg cost?

13 How many hours from noon to midnight?

14 How many 10c coins make $2.80?

15 Write 1400 in words.

____

____

16 If two apples cost 80c, what will five cost?

17 Round to estimate an answer to 217 + 31.

Statistics and Probability Column graphs

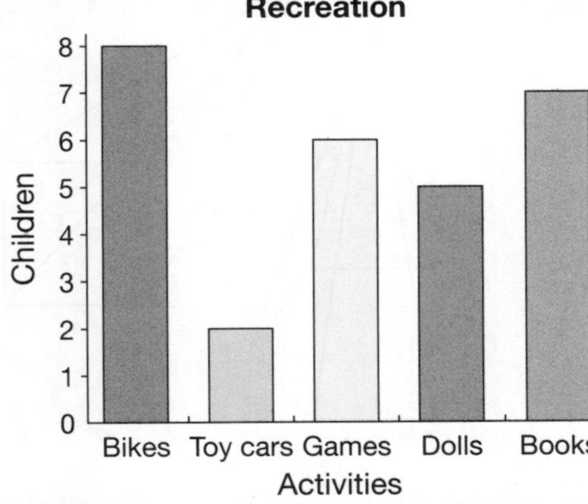

1 How many children like books best? ☐

2 How many children like bikes best? ☐

3 Which is the least popular activity? ☐

4 Which activity was chosen by 7 children? ☐

5 How many children were surveyed? ☐

6 Which 2 activities together equal bikes?

☐

UNIT 4

Number and Algebra

SET 1 Basic

1 3 + 4 + 2

2 4 + 6 + 0

3 9 – 6

4 7 – 5

5 3 × 5

6 6 × 4

7 5 × 4

8 5 × 6

9 Which is the first month of the year?

10 What is the sum of 6 and 3?

11 What is the difference between 9 and 4?

12 Add 6 and 1.

13 How many minutes in 1 hour?

14 4 tens + 6 ones

15

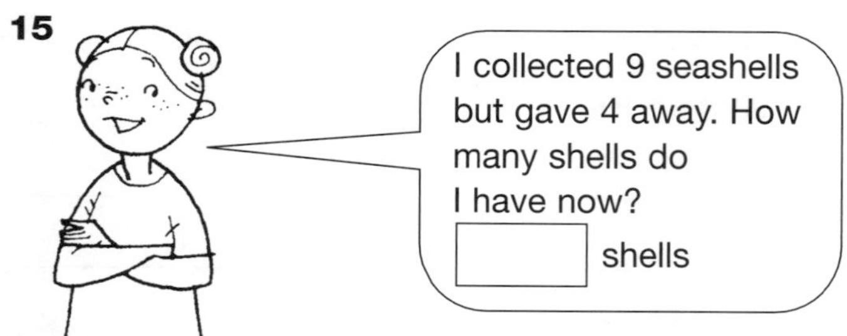

SET 2 Subtraction strategies

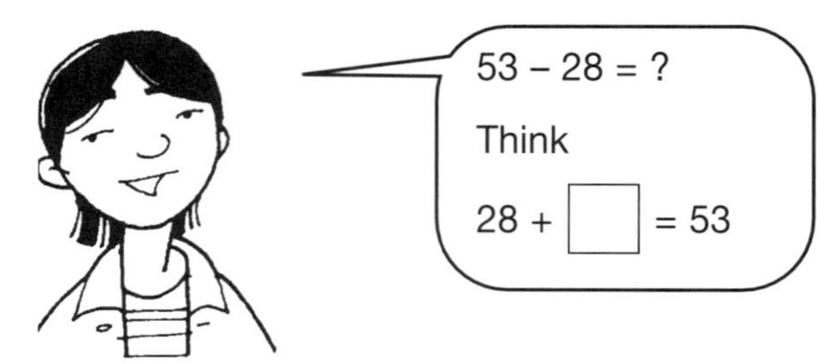

Count on to solve these subtractions.

1 43 – 21 = ☐

2 65 – 39 = ☐

3 72 – 54 = ☐

4 89 – 44 = ☐

5 32 – 17 = ☐

6 58 – 37 = ☐

7 91 – 76 = ☐

8 164 – 151 = ☐

9 88 – 55 = ☐

10 275 – 228 = ☐

11 70 take away 39

12 Subtract 56 from 98.

13 Take 25 away from 60.

Complete the grids.

14

●	86	74	49	24	13	
▲	77		40			36

15

●	35	55	85	95	125	135
▲	29		79		119	

Space Describing three-dimensional objects

Draw a line to match the description with the object.

1 I am a pyramid that has a square base and 4 triangular faces.

2 I am a pyramid with a base of 6 sides.

3 I am a pyramid with all triangular faces.

4 I have 5 triangular faces and a base of 5 sides.

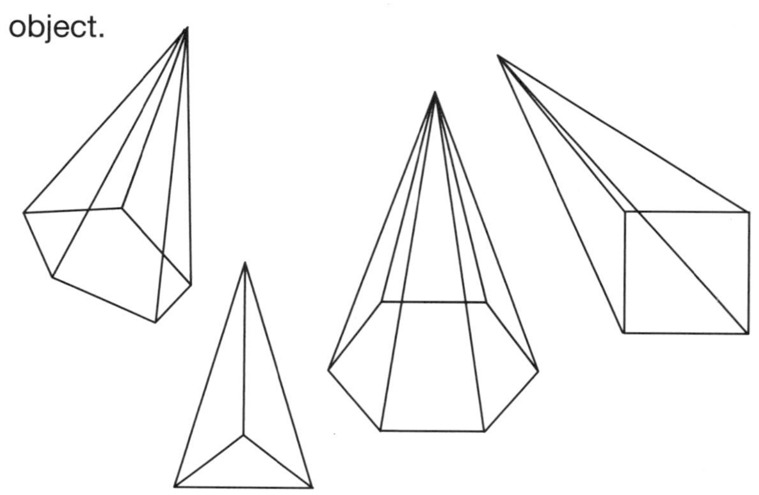

Number and Algebra

SET 3 Multiplication facts

Complete the table.

	×	3	2	5	1	0	7	9
1	3							
2	5							
3	4							
4	2							
5	10							

6 $3 × 6

7 5 kg of beans at $4 per kilo

8 6 kg of potatoes at $6 per kilo

9

How much for 4 cans? ____________

How much would 4 cans weigh? ______

How much for the entire stack? ______

Working Mathematically

SET 4 Extension

1 What is the 8th letter of the alphabet?

2 How many grams in $\frac{1}{2}$ kg?

3 Share $40 among 5 people.

4 How much is 3 kg of tomatoes at $3 per kg?

5 How much is $\frac{1}{2}$ kg of meat at $9.00 per kg?

6 What is the value of 8 in 3783?

7 How many legs on 35 people?

8 What is 13 more than 267?

9 Does a pentagon have 6 sides?

10 4 m of material costs $12. How much for 7 m?

11 Jessie threw 3 darts and scored 100. Name 3 different ways of scoring 100.

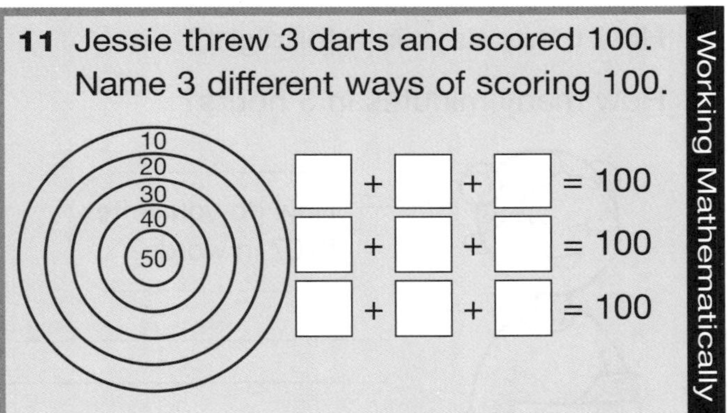

Working Mathematically

Measurement Estimating length

Estimate and measure the length of these pencils in centimetres.

1
2
3
4
5
6

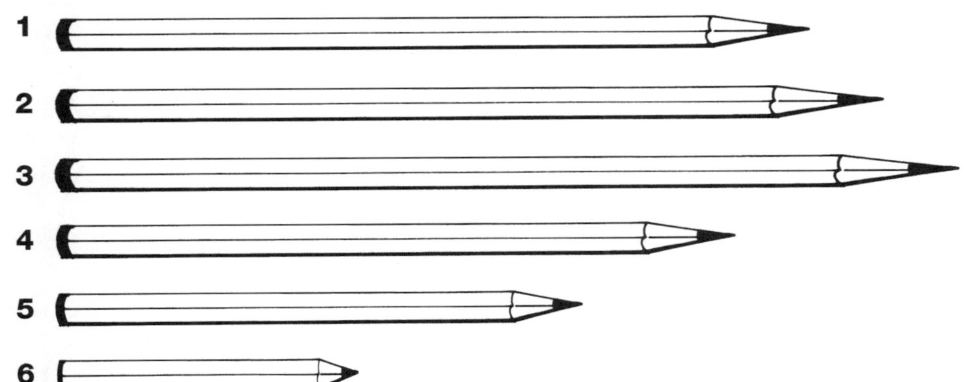

Estimate	cm

Number and Algebra

SET 1 Basic

1 5 + 6
2 7 + 3
3 18 – 4
4 25 – 5
5 3 × 6
6 4 × 2
7 5 × 4
8 6 × 3
9 What is the 6th month of the year?
10 What is the sum of 12 and 8?
11 What is the difference between 12 and 8?
12 Triple 2.
13 How many days in a fortnight?
14 How many minutes in 3 hours?
15

SET 2 Multiples of ten

Complete each pair of multiplication facts.

1	3	×	8	=	
2	3	×	80	=	
3	6	×	3	=	
4	6	×	30	=	
5	8	×	4	=	
6	8	×	40	=	
7	4	×	7	=	
8	4	×	70	=	
9	7	×	2	=	
10	7	×	20	=	
11	5	×	5	=	
12	5	×	50	=	

Space Polygons

1 Describe each shape.

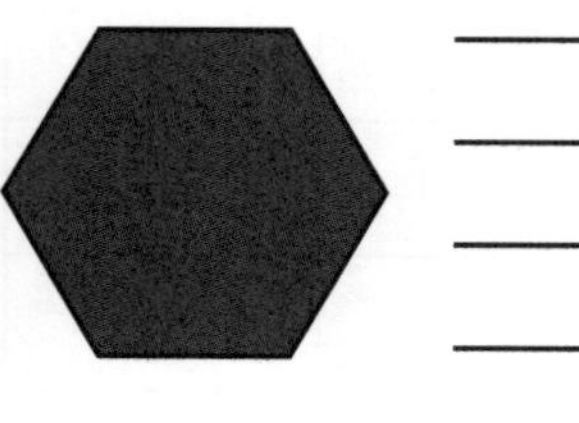

2 Colour the polygons.

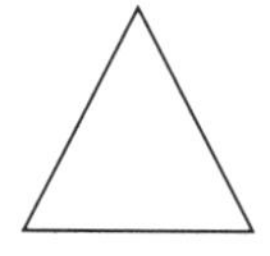

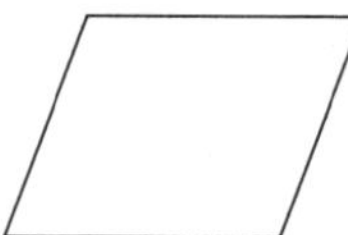
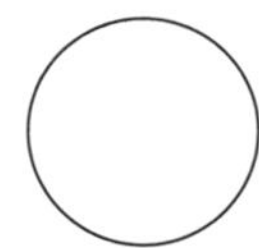
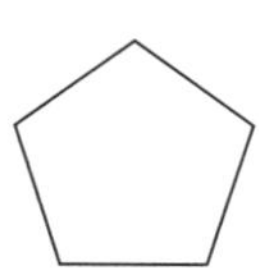
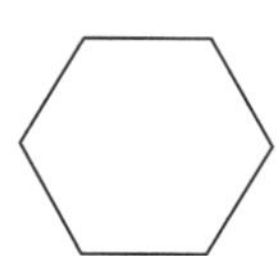
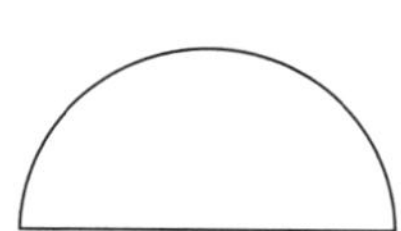
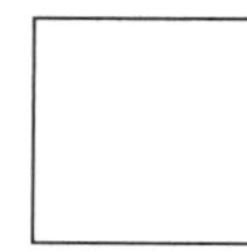
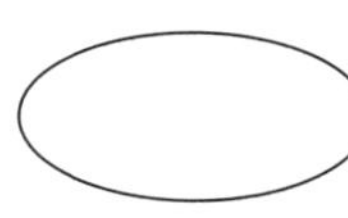

Number and Algebra

SET 3 Rounding numbers

Round each number to the nearest 10.

	Number	Rounded
1	92	
2	58	
3	179	
4	491	

Round each number to the nearest 10 to estimate the total.

5 58 + 32 becomes _____ + _____ = _____

6 73 + 28 becomes _____ + _____ = _____

7 49 + 31 becomes _____ + _____ = _____

8 72 + 37 becomes _____ + _____ = _____

9 99 + 51 becomes _____ + _____ = _____

Round each number to the nearest 10 to estimate the difference.

10 78 – 49 becomes _____ – _____ = _____

11 71 – 42 becomes _____ – _____ = _____

12 87 – 58 becomes _____ – _____ = _____

13 62 – 23 becomes _____ – _____ = _____

14 81 – 79 becomes _____ – _____ = _____

SET 4 Extension

1 (6 + 2) × 4

2 Share 36 lollies among 7 people.

3 4325 + 500

4 6 rulers at 30c each

5 What is the value of 6 in 9652?

6 What is 19 less than 36?

7 Add $0.50 to $5.60

8 How much is 9 kg of meat at $15 per kg?

9 $\frac{1}{4}$ of 24

10 How many halves in 6 wholes?

11 How much is $3\frac{1}{2}$ kg of sausages at $8 per kilogram?

12 What is the sum of 168 and 232?

13 Share $49 among 7 people.

14 402, 399, ☐, ☐, ☐, 387

Working Mathematically

15 How old is Stephen if he is $\frac{1}{2}$ the age of Anne, who is $\frac{1}{4}$ the age of Boe? Boe is 40 years old.

Measurement Time units

1 How many seconds in one minute? ☐

2 How many minutes in one hour? ☐

3 How many hours in one day? ☐

4 How many days in a week? ☐

5 How many days in a fortnight? ☐

6 How many days in a year? ☐

Select the best unit of time to record each activity (seconds, minutes, hours, days, weeks).

7 A netball game ☐

8 A marathon run ☐

9 Boiling an egg ☐

10 Putting a stamp on an envelope ☐

UNIT 6

Number and Algebra

SET 1 Basic

1 26 + 3
2 74 + 6
3 3 + 7 + 5
4 26 – 5
5 28 – 4
6 2 × 6
7 3 × 5
8 4 × 4
9 5 × 3
10 What is the product of 6 and 3?
11 19 take away 6
12 Triple 3.
13 Half of 48
14 How many hours between 2 pm and 10 pm?
15 7 thousands + 6 hundreds + 3 tens + 7 ones
16

SET 2 Numeration

1 Show 3743 on the expanders.

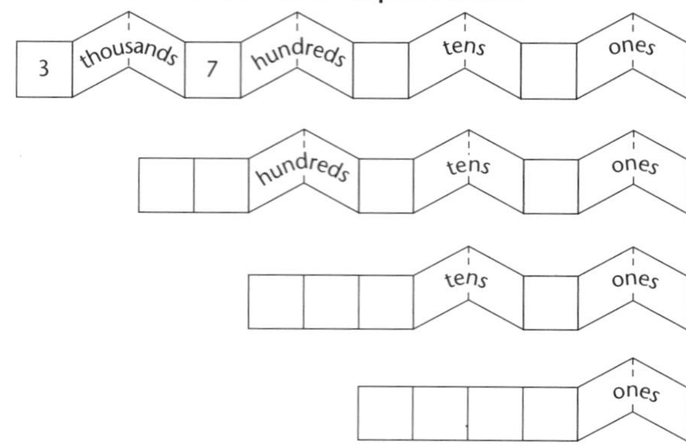

2 How many tens in 36?
3 How many tens in 137?
4 How many tens in 1246?
5 How many hundreds in 237?
6 How many hundreds in 3256?
7 How many hundreds in 2577?
8 How many thousands in 3576?
9 Add 100 to 3568.
10 Add 10 to 1357.
11 Add 10 to 8050.
12 Add 20 to 3568.
13 Add 30 to 2371.
14 Add 200 to 5870.
15 Add 300 to 1256.

Space Grid maps

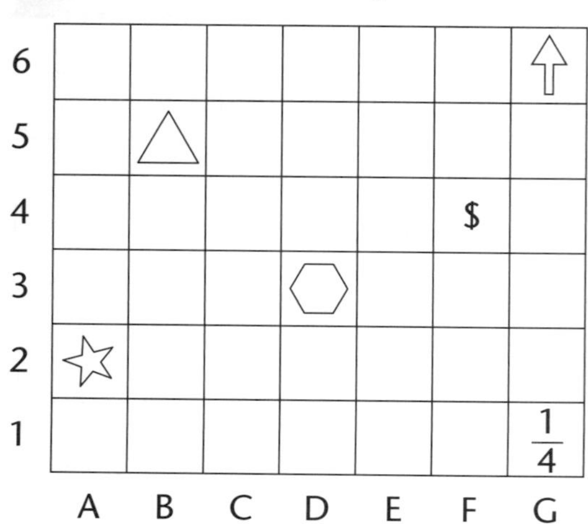

Give the grid reference points for these items.

1 ☆
2 ⬡
3 $\frac{1}{4}$
4 △
5 $
6 ⇧

7 Place a **#** at C4
8 Place a **T** at E6

Number and Algebra

SET 3 Multiplication facts, times 6

1 3 × 6 =

2 5 × 6 =

3 7 × 6 =

4 9 × 6 =

5 2 × 6 =

6 4 × 6 =

7 6 × 6 =

8 8 × 6 =

9 10 × 6 =

10 5 × 3 =

11 7 × 5 =

12 9 × 3 =

13 6 × 5 =

14 3 × 3 =

15 2 × 4 =

16 1 × 3 =

17 10 × 5 =

18 8 × 10 =

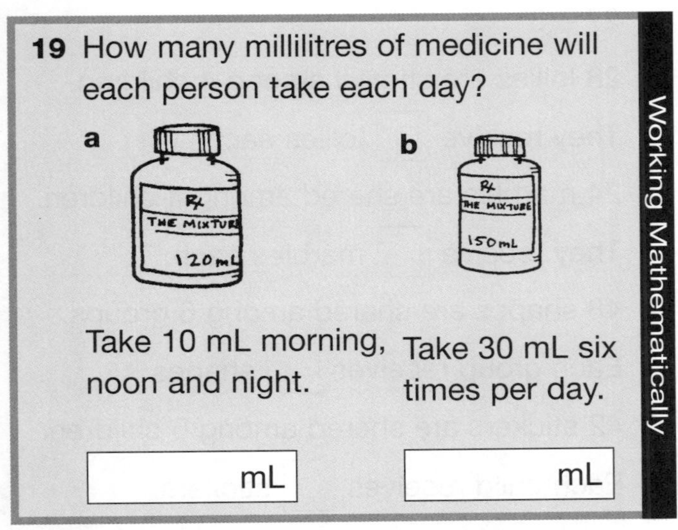

Working Mathematically

19 How many millilitres of medicine will each person take each day?

a Take 10 mL morning, noon and night. [ ] mL

b Take 30 mL six times per day. [ ] mL

SET 4 Extension

1 45 + 5 + 45

2 145 – 60

3 70 + 30

4 What is the product of 5 and 4?

5 What is the difference between 18 and 9?

6 3 lollies at 10c each

7 50 less than 1000

8 (24 – 4) + 2

9 Triple 4.

10 How many quarters in $1\frac{1}{4}$?

11 $\frac{1}{4}$ of 16

12 2340 + 300

13 Double 24.

14 4185 less one hundred

15 Write one thousand, three hundred and twenty-four in figures.

16 Complete the number web.

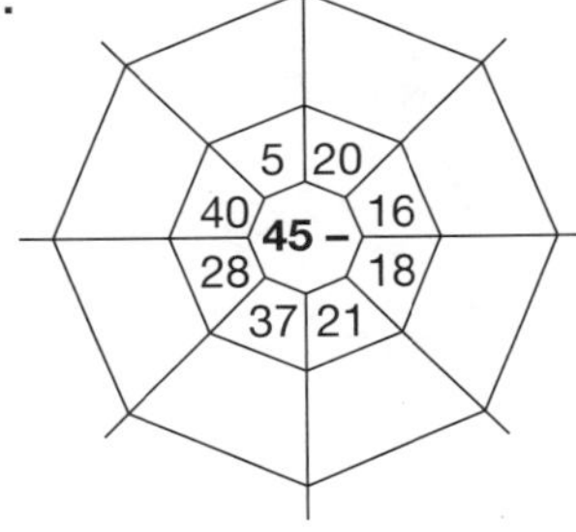

Statistics and Probability Column graphs

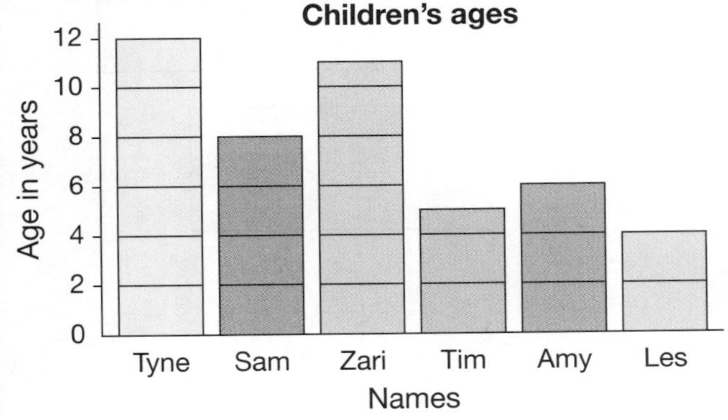

Answer the questions on the 6 children's ages.

1 How old is Les?

2 How old is Sam?

3 How old is Zari?

4 How much older is Tyne than Amy?

5 How much younger is Tim than Zari?

Number and Algebra

SET 1 Basic

1 6 + 4 + 4

2 7 + 3 + 6

3 18 – 5

4 17 – 6

5 4 × 10

6 5 × 9

7 7 × 3

8 6 × 2

9 What is the difference between 43 and 10?

10 Add 24 to 13.

11 Multiply 3 by 7.

12 What is the 4th month of the year?

13 Minutes in $\frac{1}{4}$ hour

14 (3 × 7) + 6

15

How much money have I got if I have one cheque for $3000, two $100 notes, three $10 notes and seven $1 coins?

$ ☐

SET 2 Division

1 Divide 24 by 4.

2 Divide 24 by 6.

3 Share 24 among 3.

4 Divide 32 by 4.

5 Divide 24 by 8.

6 Divide 36 by 6.

7 Share 100 among 10.

8 Share 45 among 5.

9 Share 35 among 5.

10 18 ÷ 6

11 27 ÷ 3

12 28 lollies are shared among 4 children.
They receive ☐ lollies each.

13 24 marbles are shared among 4 children.
They receive ☐ marbles each.

14 48 shapes are shared among 6 groups.
Each group receives ☐ shapes.

15 42 stickers are shared among 6 children.
Each child receives ☐ stickers.

Space Parallelograms and trapeziums

Draw a line to match each shape to a label.

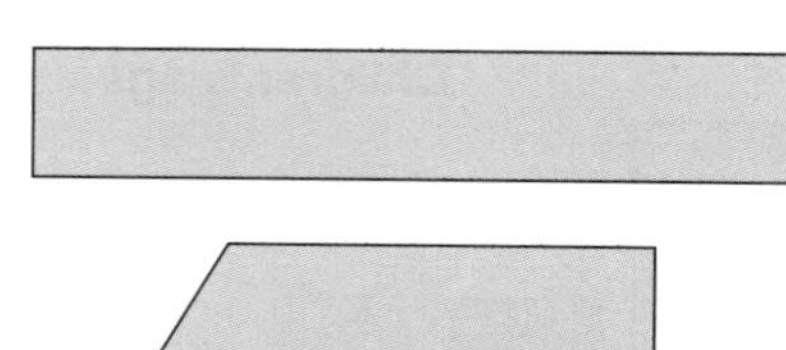

Parallelogram

Trapezium

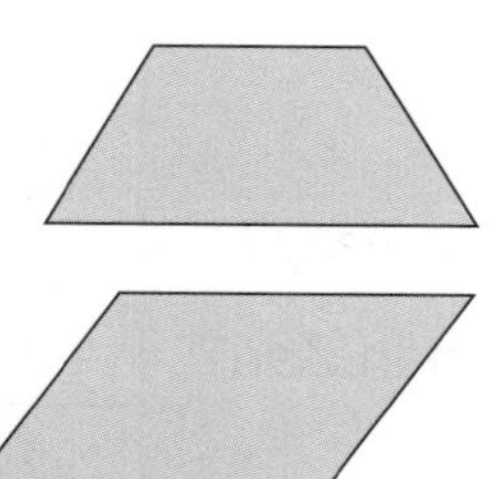

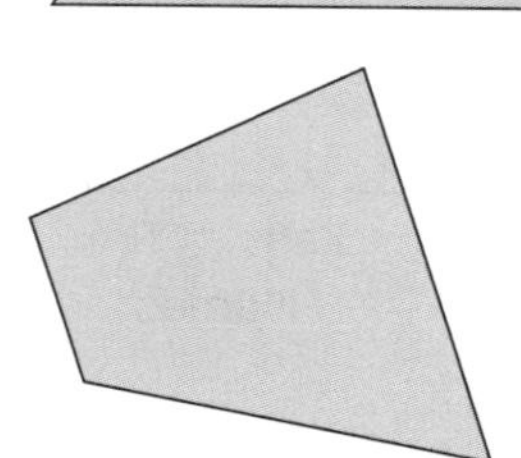

Number and Algebra

SET 3 Addition/subtraction strategies

Use any strategy you wish to answer the following.

120 + 37

= 120 + 30 + 7 = 157

1 24 + 36
2 37 + 53
3 130 + 54
4 79 – 43
5 85 – 49
6 17 + 8 + 3
7 15 + 9 + 5
8 50 – 30
9 500 – 300
10 600 – 400
11 How much space is left along a 458 cm wall if a 46 cm cupboard is placed there?

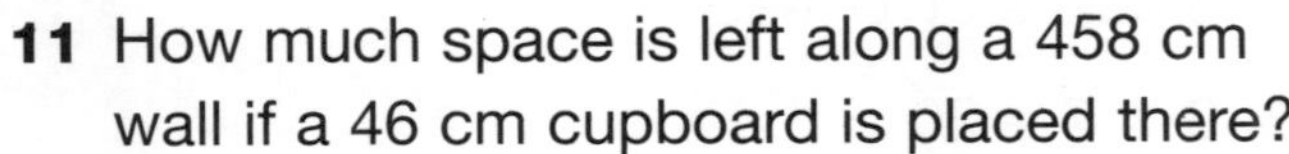

Working Mathematically

12 Place the numbers 6, 12, 18, 24 and 30 in the boxes so that the sum of the numbers is the same when added horizontally and vertically.

12

18 ☐ ☐

☐

SET 4 Extension

1 What is the total of 20, 60 and 40?
2 Subtract 19 from 350.
3 What is the value of 5 in 4517?
4 Round 2790 to the nearest 100.
5 (36 – 8) + 2
6 15 more than 149
7 Share $35 among 5 people.
8 Multiply 4 by 2 and add 6.
9 $3.15 = ☐ cents
10 $(5 \times 2) + 6$
11 $5 \times (2 + 6)$
12 How many tens in 539?

Working Mathematically

13 Hawaiian — $1 per slice; Supreme — $1.20 per slice; Special — $1.10 per slice

Name 2 orders that cost $3.30.

____

____

Measurement Litres

Match a container to each quantity.

1 1 litre ☐
2 2 litres ☐
3 4 litres ☐
4 10 litres ☐
5 20 litres ☐

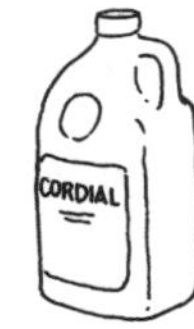

Number and Algebra

SET 1 Basic

1 4 + 3 + 6

2 7 + 2 + 1

3 19 – 3

4 21 km – 5 km

5 6 × 3

6 7 × 4

7 5 × 4

8 9 × 0

9 The 5th month of the year is ☐.

10 Sum of 15 and 4

11 Share $12 among 6 people.

12 Multiply 8 by 3.

13 Halve 20c.

14 What is the time 3 minutes after 4:15?

15 6 hundreds + 4 tens + 3 ones

16

SET 2 Multiplication facts

1

× 4	
2	
4	
5	
6	
7	
8	

2

× 7	
2	
3	
4	
6	
8	
9	

3

× 6	
2	
4	
5	
6	
9	
8	

4 6 books at $7 each

5 7 glue sticks at $3 each

6 9 presents at $8 each

7 8 cakes at $7 each

8 3 × 7 + 20

9 5 × 8 + 50

10 9 × 7 + 7

11 7 rows of 6 trees

12 9 rows of 5 plants

13 10 rows of 8 trees

14 3 × 7

15 3 × 70

Space Constructing and drawing

Each dot represents a corner of a pyramid. Draw a line between each dot and the apex dot, then draw lines between the dots for the base. Name the shape of each pyramid base.

1

2

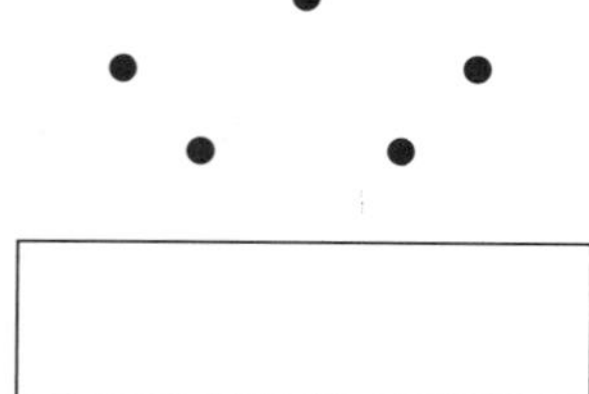

3

Number and Algebra

SET 3 Fifths and tenths

Write a fraction for each shaded shape.

1 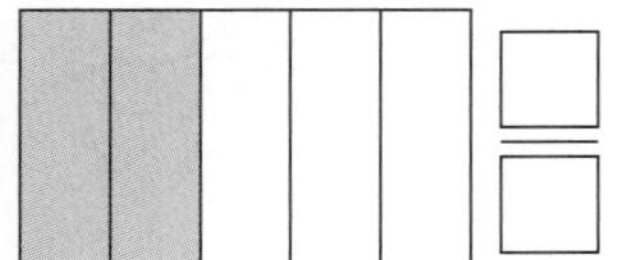

2 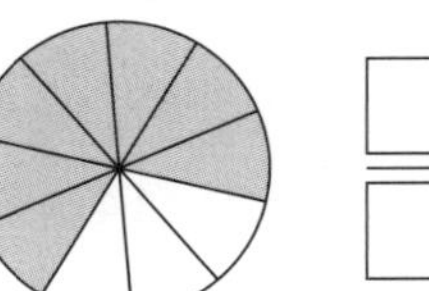

3 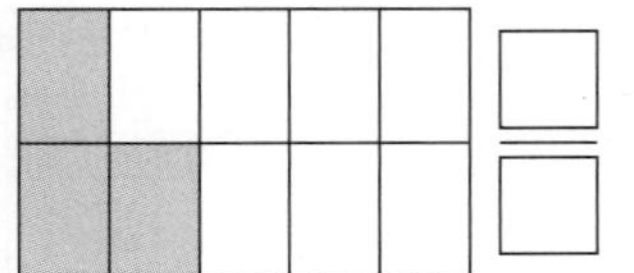

4 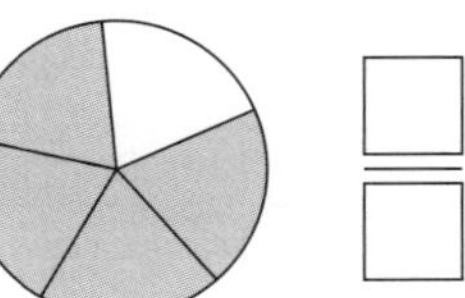

5 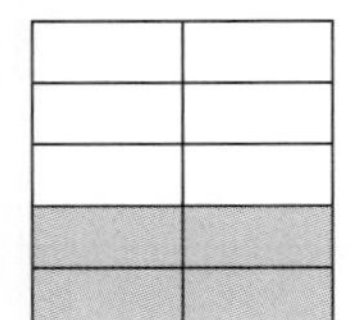

6 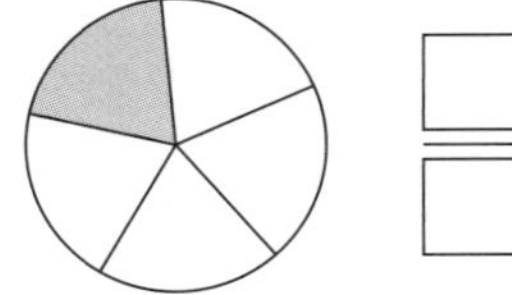

7 Order the fractions from smallest to largest: $\frac{1}{10}$, $\frac{5}{10}$, $\frac{2}{10}$, $\frac{9}{10}$.

8 How many fifths in a whole?

9 Is $\frac{4}{5}$ larger than $\frac{4}{10}$?

10 Does $\frac{3}{5} = \frac{6}{10}$?

11 Write a fraction larger than $\frac{3}{10}$.

SET 4 Extension

1 What is the 3rd letter of the alphabet?

2 Share 50c among 5 children.

3 12 + 3 + 4 + 5

4 (3 × 2) + (5 × 2)

5 1786c = $ ☐

6 What is the value of 6 in 7642?

7 Write the numeral for seven hundred and twenty-six.

8 1 fortnight + 8 days

9 If 2 kg cost $6, how much would 5 kg cost?

10 How many faces has a cube?

11 $12 – $1.50

12 $9.87 = ☐ c

13 How many hours from 3 pm to 10 pm?

14 How many legs on 4 dogs and 4 spiders?

15 Write 7260 in words.

____

____

16 How many quarters in one-half?

Measurement The square centimetre

1 Work out the area of this shape. ☐ cm^2

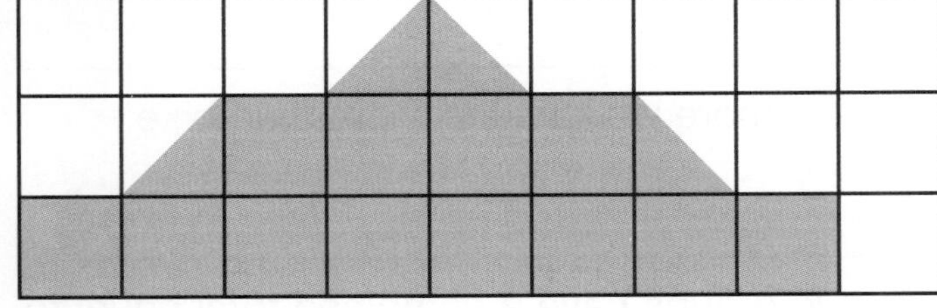

2 Draw a shape with an area of 24 cm^2.

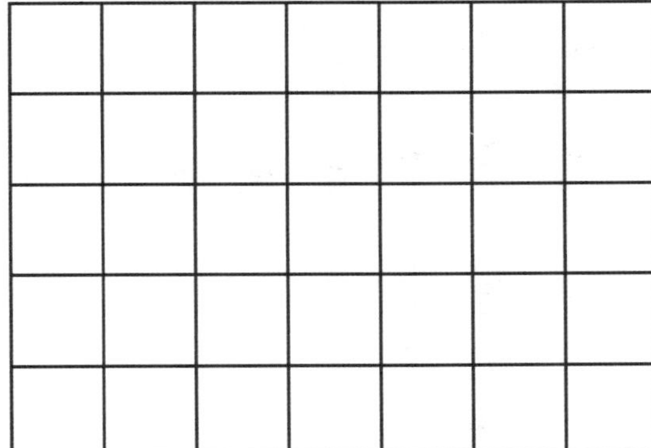

UNIT 9

Number and Algebra

SET 1 Basic

1 6 + 4 + 11

2 9 + 1 + 10

3 18 – 16

4 19 – 1

5 5 × 4

6 11 × 3

7 8 × 4

8 10 × 2

9 The 7th month of the year is ______.

10 What is the product of 6 and 8?

11 Share 8 between 2.

12 Add 4 and 21.

13 Triple 1.

14 15 minutes after 1:10 pm

15

I had $4 last month.
Now I have twice that much.
I have $ ______.

SET 2 Split strategy for addition

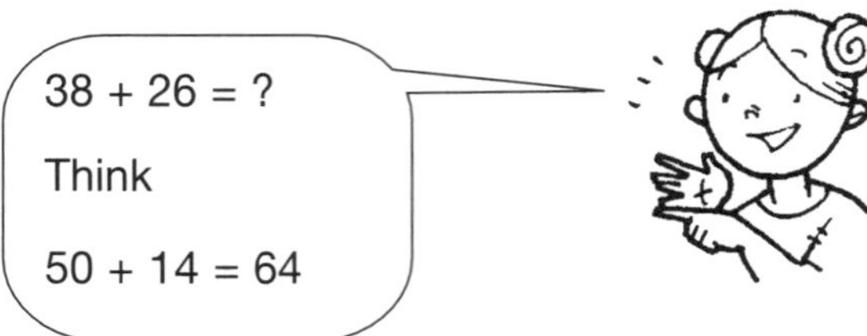

Solve these additions using the split strategy.

	Question	Becomes	Answer
1	25 + 46	60 + 11	
2	38 + 17		
3	53 + 24		
4	68 + 31		
5	47 + 55		
6	82 + 26		
7	133 + 28		
8	152 + 49		

Use the split strategy to check these additions and circle true or false.

9 62 + 39 = 101 True/False

10 26 + 43 = 79 True/False

11 55 + 37 = 92 True/False

12 112 + 67 = 179 True/False

Statistics and Probability Picture graphs

Transport to school

Bus	🕴🕴🕴🕴🕴🕴🕴 (7)
Train	(4)
Walk	(8)
Car	(5)
Bike	(3)

Number of people

1 Which is the most common form of transport?

______

2 Which is the least common form of transport?

______

3 How many more walkers are there than bike riders?

______

4 Which two groups have a combined total equal to the bus group?

______

Number and Algebra

SET 3 Multiplication strategies

To multiply by 4, double then double again.

Complete these multiplications.

1 7 × 4 = ☐

2 9 × 4 = ☐

3 11 × 4 = ☐

4 21 × 4 = ☐

5 40 × 4 = ☐

6 30 × 4 = ☐

To multiply by 5, multiply by 10 then halve.

7 14 × 5 = ☐

8 22 × 5 = ☐

9 26 × 5 = ☐

10 18 × 5 = ☐

11 46 × 5 = ☐

12 24 × 5 = ☐

Use your own strategies to solve these multiplications.

13 12 × 6 = ☐

14 11 × 5 = ☐

15 32 × 6 = ☐

16 12 × 8 = ☐

SET 4 Extension

1 How much do nine pens cost at $1.10 each?

2 What is the product of 8 and 6?

3 What is the sum of 40, 130 and 100?

4 Complete this addition sequence.

1, 3, 6, ☐, 15

5 (10 × 4) + 6

6 2295 cents = $☐

7 3 tens plus 5 tens

8 How many tens in 1486?

9 What is the difference between 89 and 65?

10 87 more than 1303

11 $\frac{1}{4}$ of 64

12 Share $1.50 among 3 people.

13 If 6 cost $3, how much would 8 cost?

14 (7 × 10) + 2 = ☐

15 Write 1964 in words.

____

____

16 Each of 4 tables has 8 people sitting at it. How many people are there altogether?

Measurement Digital time

Write the digital time for each clock face.

1

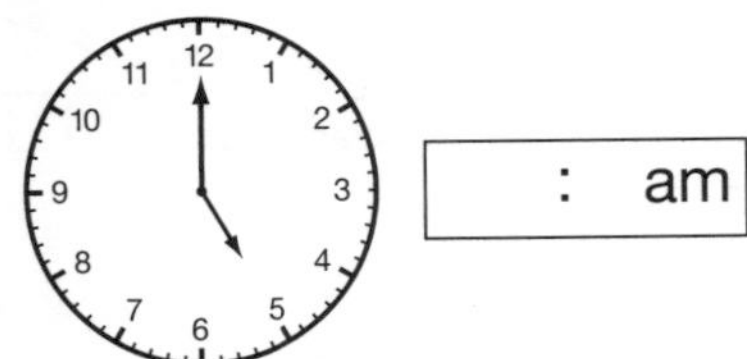
: am

2
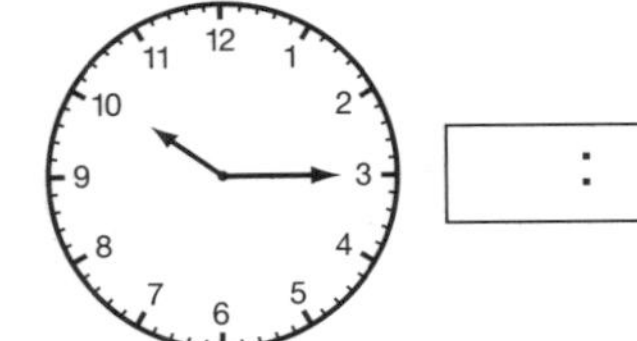
: am

3
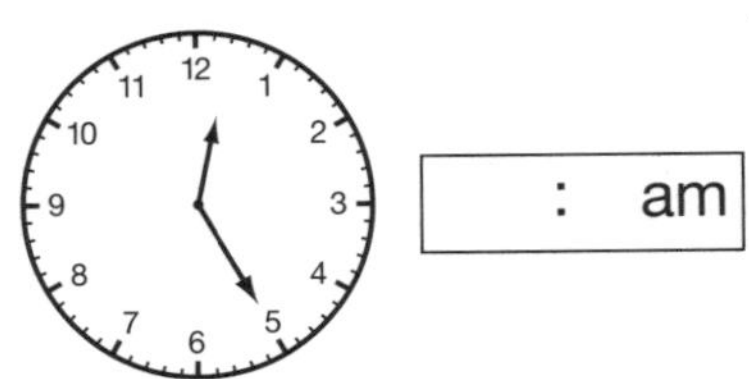
: am

4
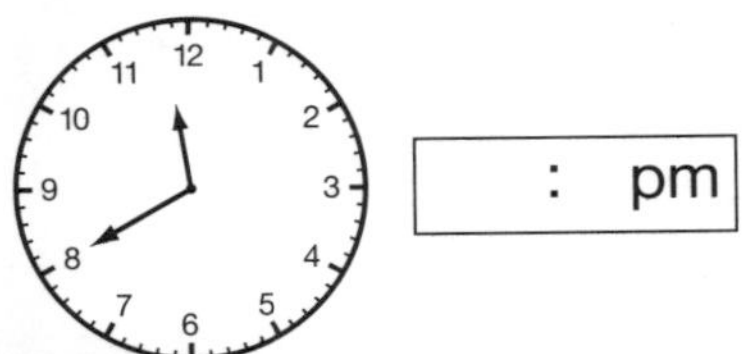
: pm

5
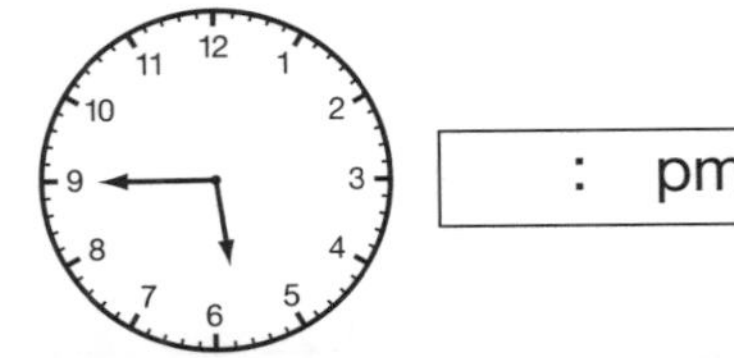
: pm

6
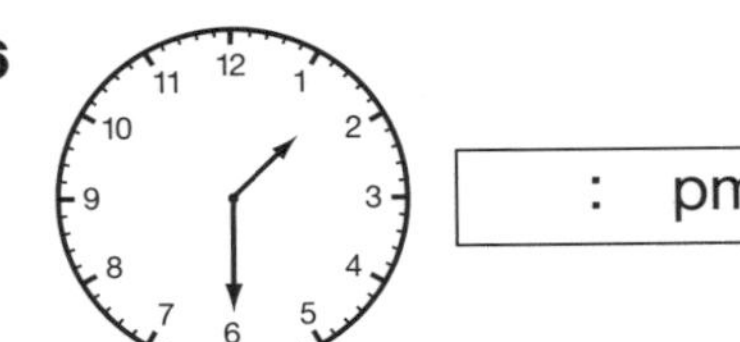
: pm

UNIT 10

Number and Algebra

SET 1 Basic

1 6 cm + 3 cm + 4 cm

2 8 – 7

3 36 – 4

4 7c + 3c + 10c

5 5 × 6

6 3 × 4

7 2 × 7

8 4 × 8

9 What is the 10th month?

10 What is the difference between 26 and 4?

11 What is the sum of 18 and 10?

12 Triple 6.

13 Double 16.

14 How many hours between 3 pm and 12 midnight?

15 7 thousands + 1 hundred + 3 tens + 7 ones

16

SET 2 Subtraction

1

	HUND	TENS	ONES
	7	7	8
–	3	3	6

2

	HUND	TENS	ONES
	7	5	3
–	5	4	1

3

	HUND	TENS	ONES
	9	2	8
–	3	0	7

4

	HUND	TENS	ONES
	6	8	9
–	4	5	8

Working Mathematically

5 Calculate the difference in distance for these plane flights from Sydney. All distances are measured in kilometres.

Destination	Distance
Newcastle	140
Port Macquarie	320
Narrabri	422
Moree	505
Melbourne	708

Narrabri	
Port Macquarie	
Difference	

Melbourne	
Moree	
Difference	

Space Angles

1 Make a list of things in your classroom that have an acute angle.

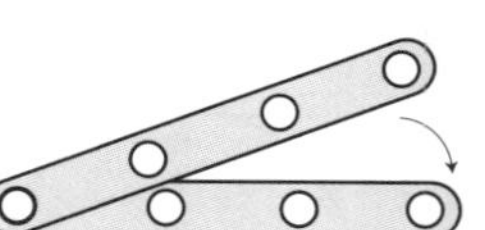

2 Make a list of things in your classroom that have an obtuse angle.

3 Make a list of things in your classroom that have a right angle.

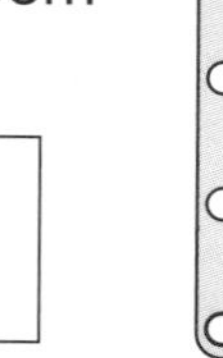

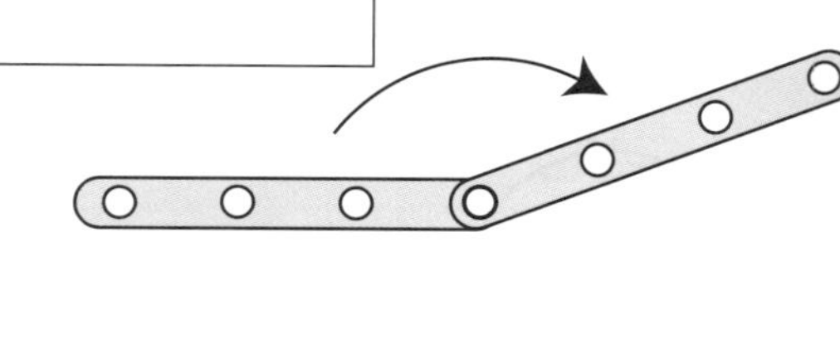

Number and Algebra

SET 3 Equivalent fractions

Shade then write an equivalent fraction for each one given.

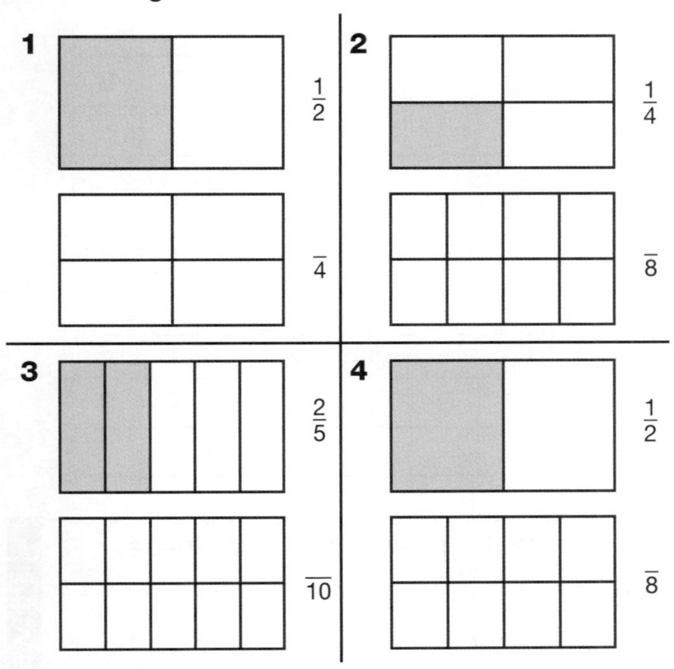

Write true or false.

5 $\frac{1}{2} = \frac{5}{10}$ ____________________

6 $\frac{1}{2} = \frac{5}{8}$ ____________________

7 $\frac{1}{4} > \frac{1}{2}$ ____________________

8 $\frac{1}{5} > \frac{1}{10}$ ____________________

9 $\frac{1}{10} < \frac{1}{2}$ ____________________

10 $\frac{7}{10} < \frac{1}{10}$ ____________________

SET 4 Extension

1 (5 × 11) + 6

2 What is the cost of 12 pens at $7 each?

3 Estimate an answer to 903 + 38.

4 9 hundreds + 6 tens + 5 ones

5 (6 × 12) + 3

6 What is the sum of 40, 30 and 250?

7 $\frac{1}{4}$ of 100

8 What is the difference between 95 and 16?

9 Write the largest number you can using 5, 2, 8, 7.

10 (6 × 8) – 4

11 How many tens in 4270?

Working Mathematically

12 Use the clues to complete the graph.

	0 4 8 12 16 20 24
Girls	(shaded to 24)
Boys	
Women	
Men	

Boys: One-half the number of girls.
Women: One-third the number of girls.
Men: One-half the number of women.

Measurement Millimetres

Measure the length of each pencil in millimetres.

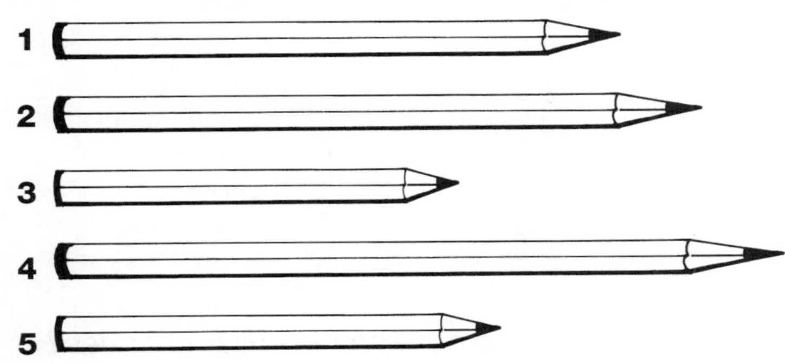

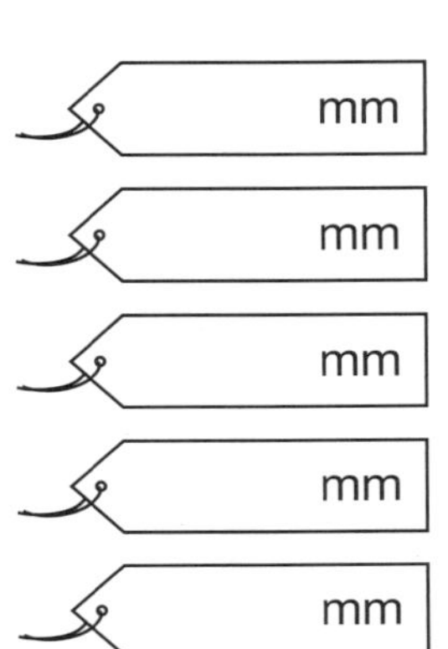

UNIT 11

Number and Algebra

SET 1 Basic

1 6 + 2 + 8

2 20 + 30 + 8

3 19 – 11

4 25 – 14

5 6 × 2

6 11 × 0

7 8 × 5

8 7 × 6

9 How many months in half a year?

10 What is the product of 5 and 9?

11 What is the sum of 5 and 9?

12 What is the difference between 5 and 9?

13 Triple 5.

14 What time is it?

15

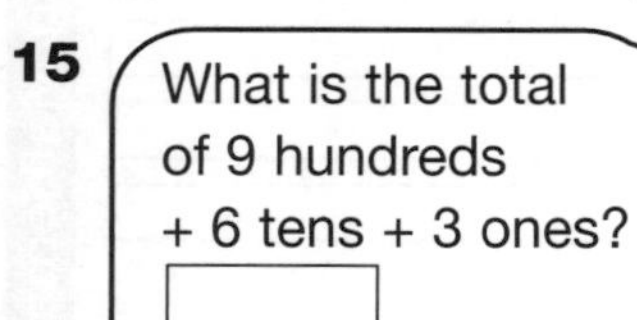

SET 2 Patterns in tables

Follow the rules to complete the number patterns.

1

Add 5							
1	2	3	4	5	6	7	8

2

Add 9							
1	2	3	4	5	6	7	8

3

Multiply by 4							
1	2	3	4	5	6	7	8

Working Mathematically

Complete the tables to answer these problems.

4 Tom walks at a rate of 4 km per hour. How far does he walk in 7 hours?

Hours	1	2	3	4	5	6	7
Km							

5 Cans of soup are packed in boxes of 8. How many cans in 7 boxes?

Boxes	1	2	3	4	5	6	7
Cans							

Statistics and Probability Chance

1 Colour the label that best describes the chance of throwing an odd number on the dice compared to throwing an even number.

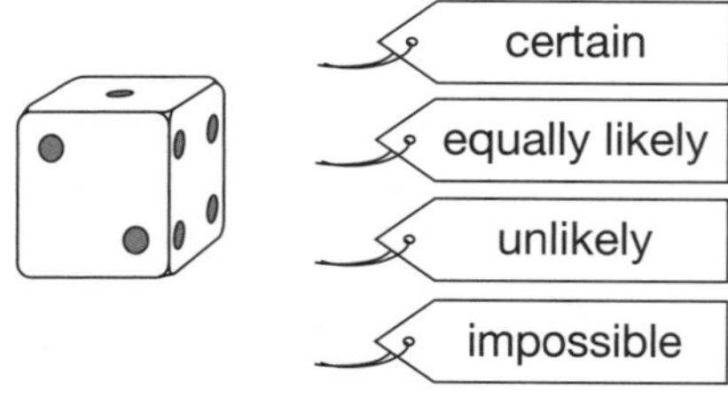

2 Colour the label that describes the chance of the spinner landing on red.

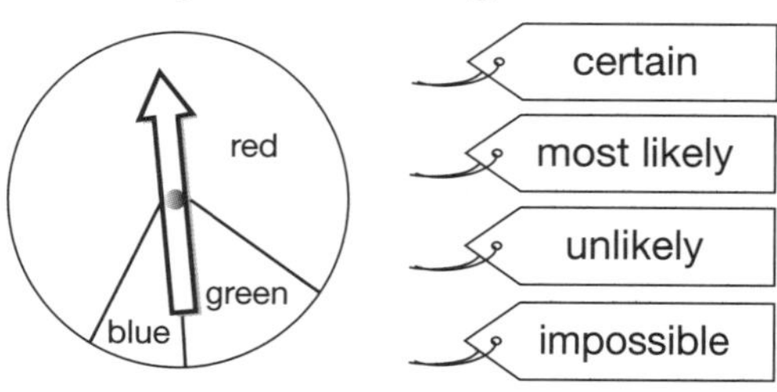

3 Which 2 colours have similar chances?

Number and Algebra

SET 3 Thirds and sixths

Shade the given fraction.

1 $\frac{1}{3}$

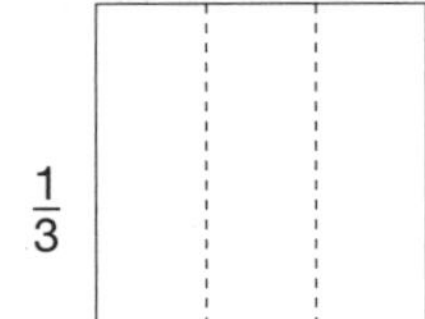

2 $\frac{2}{3}$

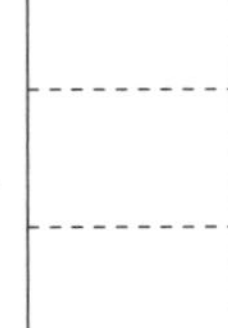

3 $\frac{2}{6}$

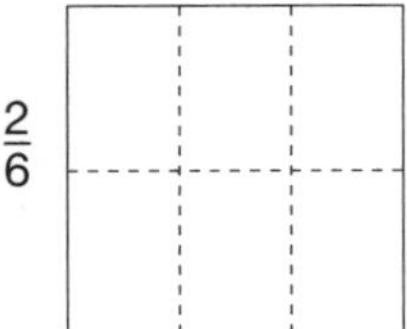

4 $\frac{4}{6}$

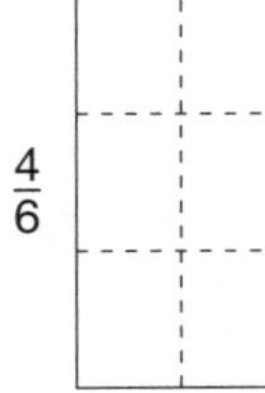

Shade the fraction of each group.

5 $\frac{1}{3}$

6 $\frac{2}{3}$

7 $\frac{1}{6}$

8 $\frac{5}{6}$

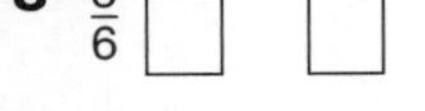

Write true or false.

9 $\frac{1}{3} = \frac{1}{10}$ = ____________

10 $\frac{1}{3} < \frac{1}{2}$ ____________

11 $\frac{1}{3} < \frac{1}{8}$ ____________

12 $\frac{1}{3} > \frac{1}{4}$ ____________

> greater than
< less than

SET 4 Extension

1. 350, 356, 362, ☐
2. 800 + 156
3. Write the largest number you can using 1, 3, 7, 6.
4. Share $100 among 4 people.
5. $\frac{1}{2}$ of 56
6. Round 7201 to the nearest 1000.
7. What is the value of 3 in 3420?
8. Write seven thousand and twenty-six in numerals.
9. How many tens in 2436?
10. How many faces does a cone have?
11. How many minutes are there from 3 pm to 7:30 pm?
12. How much is 8 kg of vegetables at $7 per kilogram?
13. How much is $4\frac{1}{2}$ kg of meat at $8 per kilogram?

Working Mathematically

14. 8 people were supposed to share the $80 hire fee for a tennis court but Prani forgot her money. How much did Carly pay if she paid her share as well as Prani's share?

Measurement Mass in kilograms

Write the item that is heavier.

1. A large bucket full of water **or** a 1 kg mass
2. A TV set **or** a 1 kg mass
3. A pair of sunglasses **or** a 1 kg mass
4. A 2 kg bag of potatoes **or** a 1 kg mass
5. This book **or** a 1 kg mass
6. A 1 kg bag of feathers **or** a $\frac{1}{2}$ kg bag of rocks
7. A 995 g mass **or** a 1 kg mass

Number and Algebra

SET 1 Basic

1 8 + 4 + 3

2 8 + 8 + 8

3 24 – 10

4 35 – 6

5 7 + 7 + 7

6 4 + 4 + 4

7 10 × 7

8 4 × 9c

9 14 kg + 5 kg

10 18 + 4

11 5 + ☐ = 20

12 31 – 1

13 25 – 6

14 16 – ☐ = 12

15 6 hundreds + 3 tens + 9 ones

16

Rulers 60c	Pens $2	Pencils 40c

How much are 4 pens, please? ☐

SET 2 Counting patterns

Complete the counting patterns.

1	64	66	68	70				
2	34	38	42	46				
3	25	30	35	40				
4	130	120	110	100				
5	250	270	290	310				
6	46	52	58	64				
7	450	470	490	510				

Follow the rules to complete the patterns.

8

Add 13							
1	2	3	4	5	6	7	8
			17				21

9

Multiply by 4							
1	2	3	4	5	6	7	8
		12					

10

Add 2 then subtract 3							
2	4	6	8	10	12	14	16
1		5					

Space Symmetrical patterns

Complete the symmetrical pattern.

Number and Algebra

SET 3 Multiplication facts, times 8

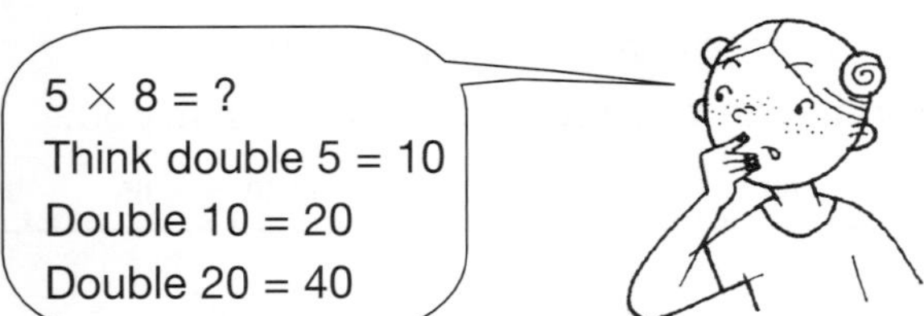

Use the double, double, double again strategy to solve these questions.

1 $3 \times 8 =$ ☐

2 $5 \times 8 =$ ☐

3 $7 \times 8 =$ ☐

4 $9 \times 8 =$ ☐

5 $10 \times 8 =$ ☐

6 $6 \times 8 =$ ☐

7 $4 \times 8 =$ ☐

8 $11 \times 8 =$ ☐

9 $2 \times 8 =$ ☐

10 $12 \times 8 =$ ☐

Complete the multiplication targets.

11

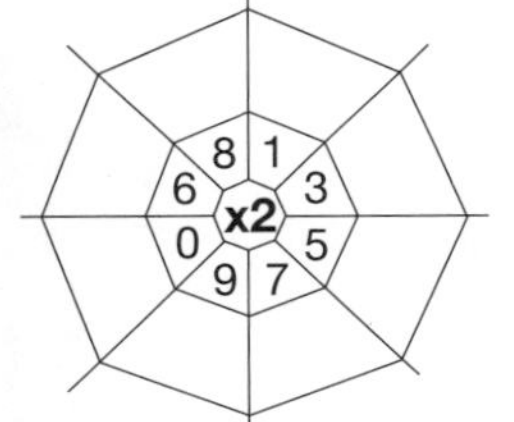

12

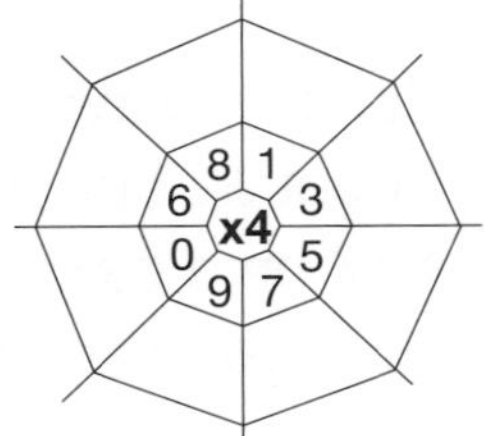

13

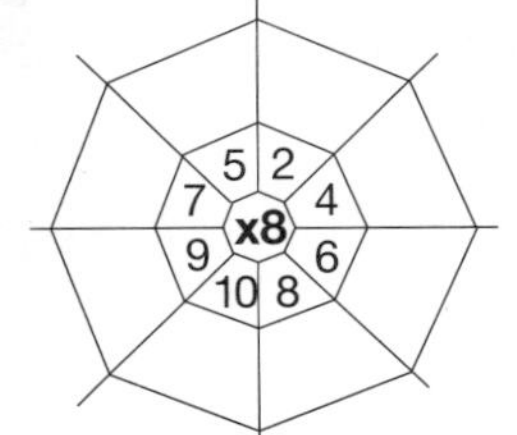

14

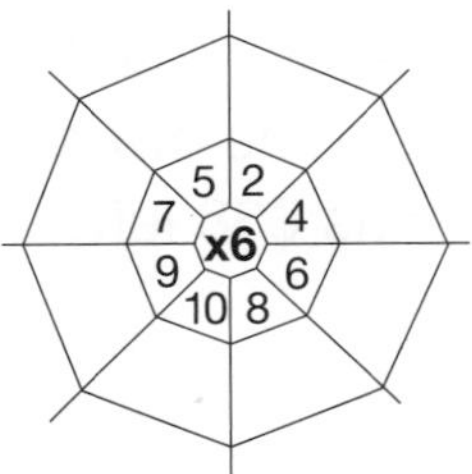

SET 4 Extension

1 $(2 \times 4) + 10$

2 39 more than 261

3 Halfway between 40 and 50

4 Estimate an answer to 1401 + 59.

5 $1.64 + $5.00

6 375, 390, 405, ☐, ☐

7 Two positions before 35th.

8 How many legs on 3 chairs with 3 people on them?

9 Share $85 among 5 people.

10 If 4 shirts cost $80, how much would 10 cost?

11 What is the value of 4 in 6400?

12 2186 cents = $ ☐

13 399 take away 2 tens

14 Subtract (2×10) from 600.

15 What is half of 84?

16 $(18 \div 3) + 87$

17 Find the quotient of 132 and 11.

Statistics and Probability Probability

1 Where do you think you will swim next? Connect each situation to its likelihood.

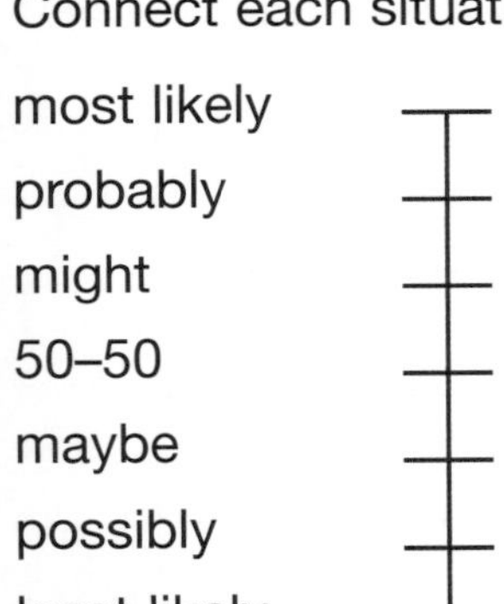

- beach
- public pool
- river
- lake
- private pool

2 What is the likelihood of the hairdresser's shop being open now? Match each sign to a chance card.

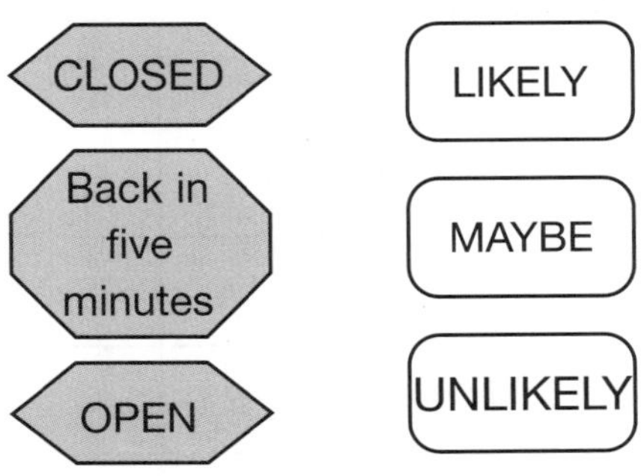

Number and Algebra

SET 1 Basic

1 12 – 3

2 16 – 4

3 8 + 20

4 23 + 7

5 4 × 8

6 6 × 8

7 3 × 8

8 7c + 7c + 7c

9 Subtract 6 from 37.

10 $3.86 = ☐ c

11 How many 5c coins in 35c?

12 How many threes in 15?

13 What is the sum of 16 and 9?

14 1376 = ☐ thousand + ☐ hundreds + ☐ tens + ☐ ones

15 

How do you write nine hundred and twenty-six as a number? ☐

SET 2 Subtraction strategies

Use any strategy you wish to answer the questions. The number line may assist you.

50 55 60 65 70 75 80 85 90

1 88 – 35

2 79 – 26

3 81 – 26

4 72 – 35

5 83 – 34

6 84 – 39

7 74 – 31

8 87 – 42

9 149 – 35

10 Jackson saved $84 but spent $20 on a book and $17 on a calculator. How much does he have left? 

11 Mary had $79 but spent some on a dress. How much did the dress cost if she has $38 left? 

Space Combining 2D shapes

How many letters of the alphabet can you make using 5 squares? Use the grid paper to display your answer.

Number and Algebra

SET 3 Mixed numerals

Colour the shapes to match the mixed numerals.

1 $1\frac{2}{5}$

2 $2\frac{3}{8}$

3 $1\frac{7}{10}$

4 $3\frac{1}{4}$

Write the next mixed numeral in the sequence.

5 1, $1\frac{1}{4}$, $1\frac{2}{4}$, ☐

6 $3\frac{1}{5}$, $3\frac{2}{5}$, $3\frac{3}{5}$, ☐

7 2, $2\frac{1}{2}$, 3, ☐

8 $1\frac{5}{10}$, $1\frac{6}{10}$, $1\frac{7}{10}$, ☐

SET 4 Extension

1 Share 70c between 2 people.

2 How many 10c coins make $4.00?

3 How much is 4 kg of onions at 55c per kg?

4 What is the value of 9 in 5945?

5 Estimate an answer to 1596 + 201.

6 Round 1594 to the nearest 10.

7 Round 1594 to the nearest 100.

8 Round 2680 to the nearest 1000.

9 How many hours are there from 6 am to 6 pm?

10 Multiply 6 by 4 and add 5.

11 9, ☐, 27, ☐, ☐, 54, ☐

12 What is the change from $3 if I spent $1.65?

13 Make the smallest number possible using 9, 6, 9, 6.

Working Mathematically

14 How much money did the children pay for their excursion? On Monday their teacher collected $40, on Tuesday she collected twice as much and on Wednesday she collected $120.

Measurement Millilitres

1 How many millilitres in 1 L?

2 How many millilitres in $\frac{1}{2}$ L?

3 How many millilitres in 4 L?

4 How many millilitres in $\frac{1}{4}$ L?

5 How many millilitres in $3\frac{1}{2}$ L?

6 Five litres equals how many millilitres?

7 How much is left in a half-litre bottle if 250 mL is poured out?

8 I have half a litre of milk. How many more millilitres do I need to make 1 L?

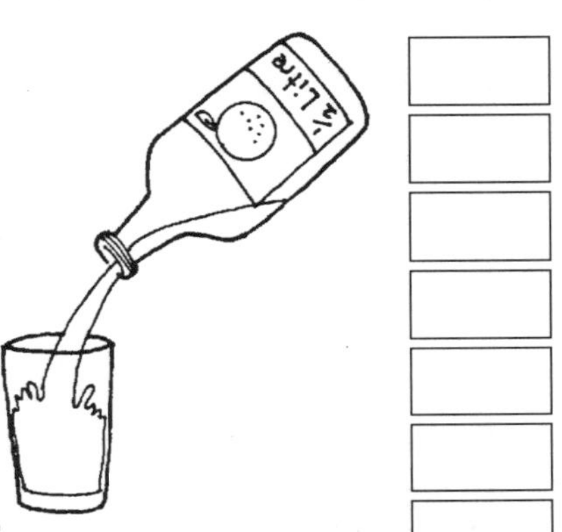

UNIT 14

Number and Algebra

SET 1 Basic

1 15 + 7

2 4 + 8 + 8

3 21 – 9

4 37 cm – 5 cm

5 7×3

6 5×3

7 $(4 \times 4) + 5$

8 Five lollies at 10c each

9 18 kg + 7 kg

10 Half of 30

11 How many days in 3 weeks?

12 Triple 8.

13 How many hours between 1 am and midday?

14 What is the difference between 33 and 5?

15 How much do 2 tickets to 'Ghost Dusters' cost?

ADMIT ONE
GHOST DUSTERS $5.50

SET 2 Counting with fractions

Continue each sequence.

1

0	$\frac{1}{4}$	$\frac{2}{4}$	$\frac{3}{4}$	1	$\frac{5}{4}$		

2

0	$\frac{1}{2}$	1	$1\frac{1}{2}$	2	$2\frac{1}{2}$		

3

0	$\frac{1}{5}$	$\frac{2}{5}$	$\frac{3}{5}$	$\frac{4}{5}$		

4

0	$\frac{1}{8}$	$\frac{2}{8}$	$\frac{3}{8}$	$\frac{4}{8}$	$\frac{5}{8}$	$\frac{6}{8}$	$\frac{7}{8}$	1		

Place these fractions and mixed numerals in ascending order.

5 $\frac{3}{10}, \frac{1}{10}, \frac{7}{10}$ ____________

6 $\frac{3}{4}, \frac{1}{4}, \frac{2}{4}$ ____________

7 $\frac{7}{8}, \frac{3}{8}, \frac{5}{8}$ ____________

8 $1\frac{1}{4}, 1\frac{3}{4}, 1\frac{2}{4}$ ____________

9 $2\frac{3}{5}, 2\frac{1}{5}, 2\frac{2}{5}$ ____________

True or false?

10 $1\frac{7}{10} < 1\frac{9}{10}$ ____________

11 $2\frac{1}{4} > 2\frac{3}{4}$ ____________

12 $2\frac{1}{2} > 2\frac{1}{4}$ ____________

Space Translate, rotate and reflect

Draw the result if each shape is translated, reflected and rotated.

Object	Translate	Reflect	Rotate

Number and Algebra

SET 3 Division strategies

Use halving skills to solve the divisions.

1 40 ÷ 2

2 18 ÷ 2

3 Divide 36 by 2.

4 Divide 64 by 2.

Use the halve and halve again strategy to solve these divisions.

5 24 ÷ 4

6 40 ÷ 4

7 Divide 32 by 4.

8 Share 20 lollies among 4 people.

Use the array to answer questions 9 to 13.

9 How many 5s in 20?

10 How many 4s in 20?

11 20 ÷ 2

12 20 ÷ 1

13 ☐ × 5 = 20

SET 4 Extension

1 What is the 5th letter of the alphabet?

2 How many times can $4 be taken away from $20?

3 Share $1.00 among 5 children.

4 18c + 17c + 19c + 13c

5 How many days altogether in January and November?

6 What is the value of 6 in 6321?

7 Estimate an answer to 999 + 496.

8 How much is 3 kg of sugar at $1.50 per kg?

9 Round 3921 to the nearest 1000.

10 Is a cylinder a 2D object?

11 How many 5c coins make $1?

12 10 books at $17 each

13 How many minutes between 1 pm and 4 pm?

14 Seven lots of five

15 Write 3062 in words.

____

____

16 Each player receives 15 cards. How many cards do I need for 3 players?

Statistics and Probability Representing data

Make a column graph to show the length of the pencils.

Pencils	Tally
19 cm	||||
18 cm	卌
17 cm	卌 ||
16 cm	卌
15 cm	|||
14 cm	卌
13 cm	||

Length of pencils

Length of pencils								
19 cm	▒	▒	▒	▒				
18 cm								
17 cm								
16 cm								
15 cm								
14 cm								
13 cm								
	1	2	3	4	5	6	7	8

Number of pencils

UNIT 15

Number and Algebra

SET 1 Basic

1 18 + 4

2 6 × 7

3 (5 × 3) – 1

4 $6 × 4

5 3^2

6 4 + 4 + 4 + 4

7 How much are 8 lollies at 5c each?

8 91 less 3

9 64 kg – 9 kg

10 What is the product of 11 and 4?

11 What is the difference between 21 and 3?

12 Half of 4 × 5

13 Subtract 6 from 21.

14 3:30 pm + 25 mins

15 ☐ hundreds + ☐ tens + ☐ ones = 695

16

How much change would I receive from $1 if I spent 85c?

☐ c

SET 2 Fractions and decimals

Record the shaded part of each shape as tenths and decimals.

1

$\frac{\square}{10}$ ☐ . ☐

2

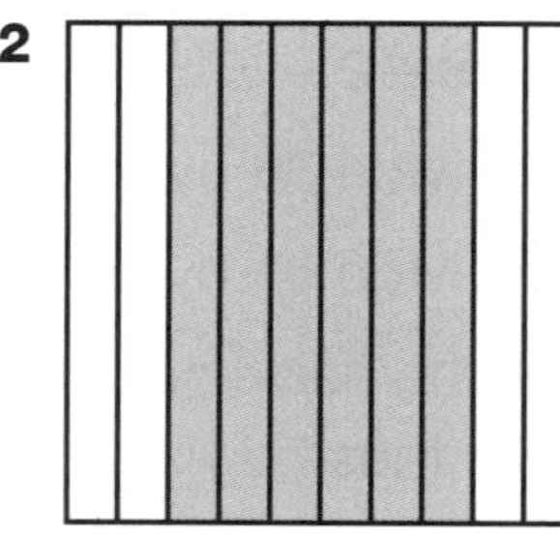

$\frac{\square}{10}$ ☐ . ☐

3

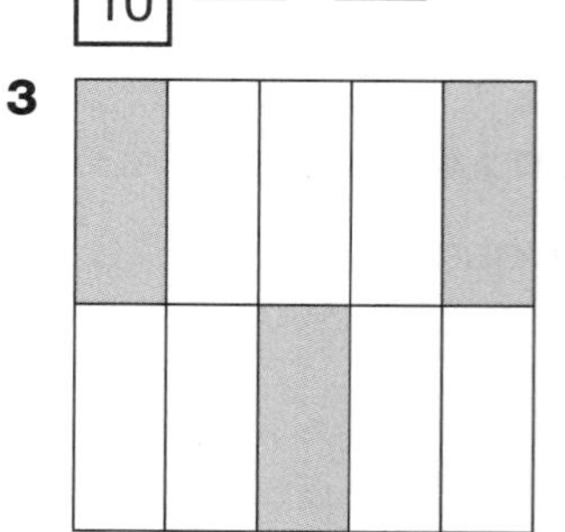

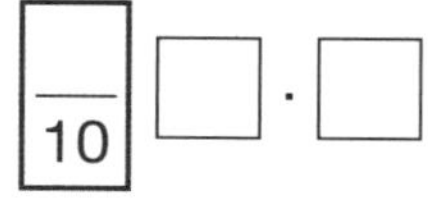

4

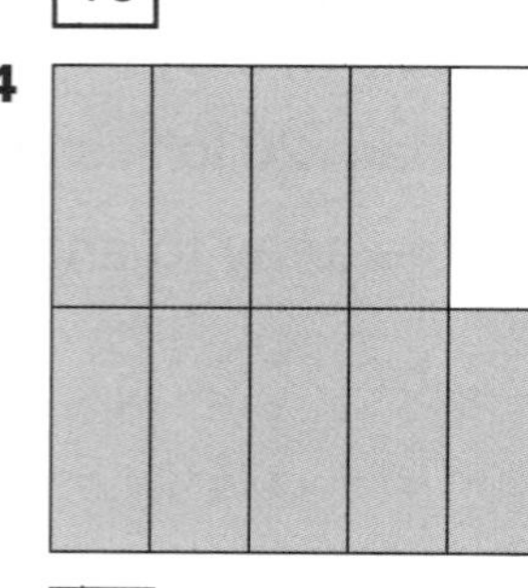

5 Connect each decimal to a point on the ruler.

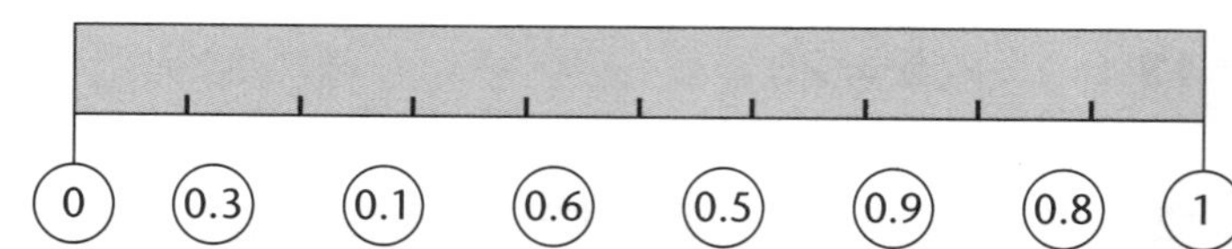

Space Grouping two-dimensional shapes

1 Draw two different shapes with angles less than a right angle. For example:

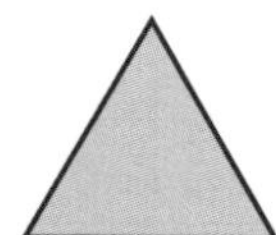

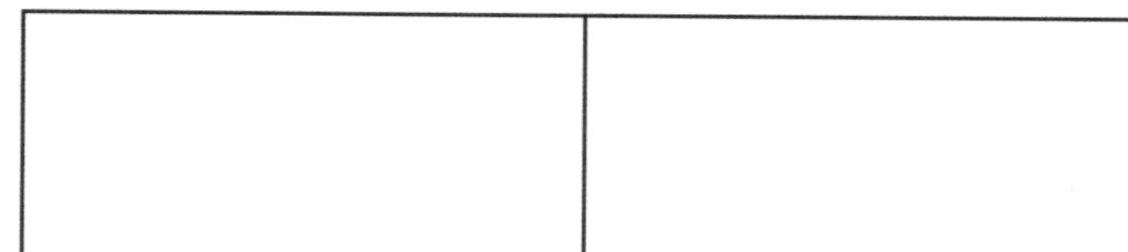

2 Draw two different shapes with parallel sides. For example:

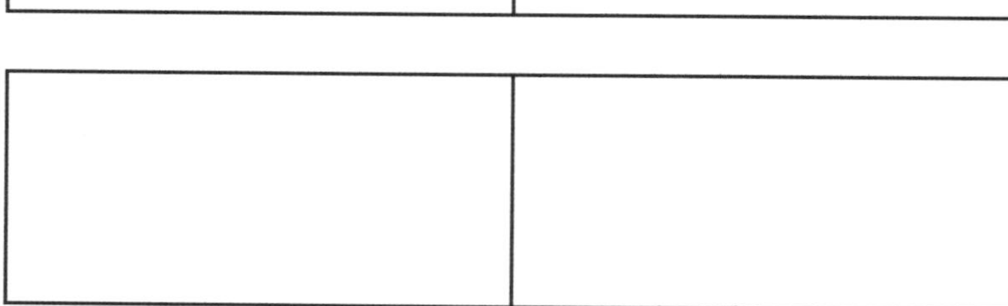

Number and Algebra

SET 3 Place value

tens hundreds thousands ones tens of thousands

Use the words in the box to write the place value of the bold numbers.

1	35**7**72	hundreds
2	374**5**6	
3	3567**8**	
4	5**6**434	
5	98**5**53	
6	**2**3657	
7	4**4**538	

Write these numbers.

8 Twenty-three thousand, two hundred and ninety-six __________

9 Thirty-five thousand, four hundred and thirteen __________

10 Fifty-four thousand and eighty-six __________

SET 4 Extension

1 Add $6 to $17.80.

2 Subtract $9 from $13.60.

3 5 pens at $1.95 each

4 19 places behind 35th

5 What number is halfway between 38 and 46?

6 18, 36, ☐, ☐, 90, ☐

7 If 4 rulers cost 88c, how much would 6 cost?

8 Change from $5 if I spent $3.45

9 List the factors of 36.

10 How many centimetres in $3\frac{1}{2}$ m?

11

Tathra		Green Point	
Points	69	Points	95
Penalties	11	Penalties	7

Which team won and what was the difference in their point scores?

12 Round to the nearest 10 to find an answer for 427 + 153.

13 ☐ × (5 + 6) = 99

Measurement Analog clocks

Draw these times on the clock faces.

1 9 o'clock

2 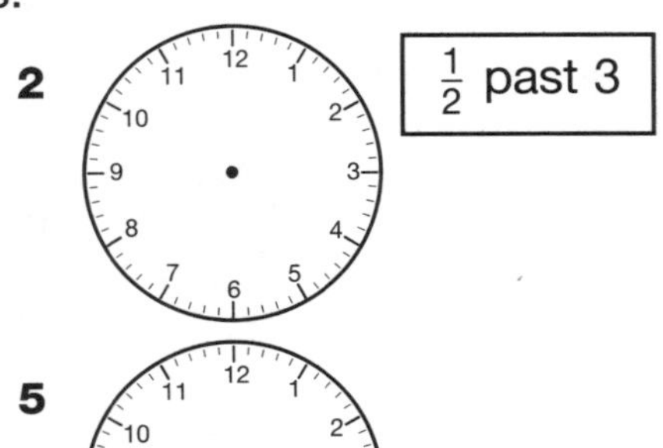 $\frac{1}{2}$ past 3

3 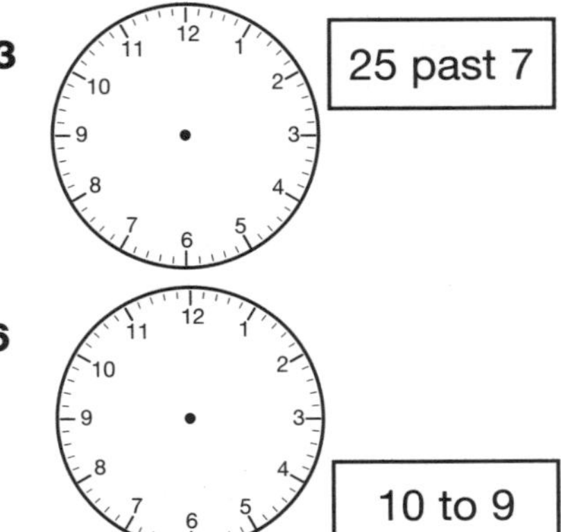 25 past 7

4 $\frac{1}{2}$ past 10

5 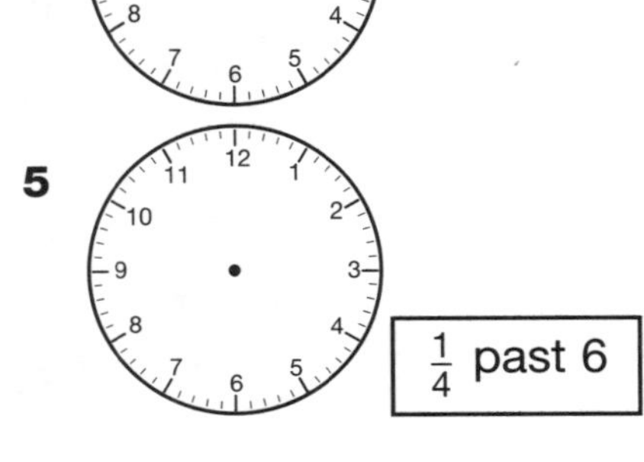 $\frac{1}{4}$ past 6

6 10 to 9

UNIT 16

Number and Algebra

SET 1 Basic

1 47 − 5

2 42 − 5

3 26 + 7

4 36 + 7

5 7×4

6 6×3

7 5×7

8 8×6

9 7 + 3 + 12

10 8 + 7 − 5

11 What is the difference between 20 and 6?

12 How many fours in 16?

13 What is the sum of 52 and 11?

14 What is the product of 9 and 2?

15 How much will it cost Tran to buy her lunch?

Item	Cost
1 sandwich	$1.20
1 drink	$0.80
1 apple	$0.60
Total	

SET 2 Addition

1

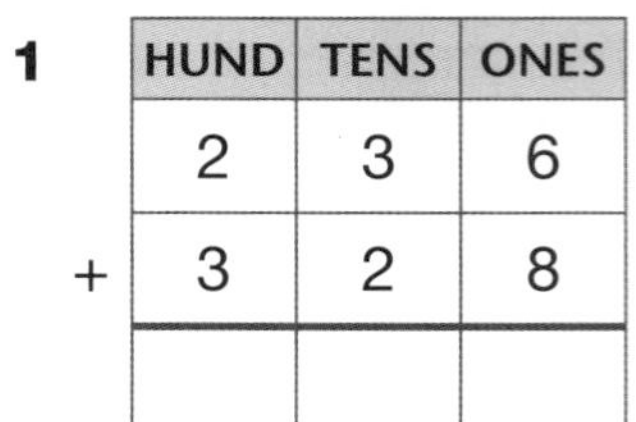

	HUND	TENS	ONES
	2	3	6
+	3	2	8

2

	HUND	TENS	ONES
	3	4	7
+	4	2	6

3

	HUND	TENS	ONES
	6	2	6
+	1	4	8

4

	HUND	TENS	ONES
	8	3	7
+		2	4

5

	HUND	TENS	ONES
	3	5	6
+	5	3	4

6

	HUND	TENS	ONES
	5	2	6
+	2	2	6

7 Add 25 to 168.

8 What is 28 more than 347?

9 What is the sum of 65 and 102?

10 What is the total of 510 and 69?

11 How much will it cost Cheyenne to buy both these rings?

 $178

 $244

Statistics and Probability Organising data by classifying

Organise these items into the three categories.

Fruit

Toys

Jewellery

Number and Algebra

SET 3 Multiplication facts, times 9

Complete the table.

	Numbers	×9	×3	×6
1	3			
2	7			
3	10			
4	2			
5	6			
6	5			
7	8			

8 A boy bought 9 books at $8 each. How much did he spend?

9 A minibus that carries 9 passengers took Mr King's class to the beach. If the minibus made 9 trips, how many children are in the class?

10 Nine packets of toilet paper were delivered to the school. Each packet contained 10 rolls. How many rolls of toilet paper were delivered?

SET 4 Extension

1 Write the largest number you can using 2, 0, 9, 6.

2 How many sides on 5 pentagons?

3 How many minutes in $3\frac{1}{2}$ hours?

4 Share $28 among 4 people.

5 331, 340, 349, ☐, ☐, 376

6 If 6 hats cost $30, how much do 8 cost?

7 Add $1.50 to $4.25.

8 21 places behind 415th

9 How many halves in 8 wholes?

10 List the factors of 24.

11 How much does 3 kg of meat cost at $16 per kg?

12 How many legs on 4 bees and 7 dogs?

13 What is half of 660?

Working Mathematically

14 Jennifer is trying to solve 8 × 6 on her calculator but the 6 button is missing.

Place numbers in the boxes so that her problem is solved.

8 × ☐ × ☐ = 48

Measurement Grams

Find the difference in mass between:

1 a can of baked beans and a can of apples. ☐

2 a jar of jam and a can of baked beans. ☐

3 a bag of sweets and a packet of tea. ☐

4 a jar of jam and a can of apples. ☐

5 a jar of jam and a bag of sweets. ☐

UNIT 17

Number and Algebra

SET 1 Basic

1 40 + 7

2 83 + 6

3 27 – 5

4 45c – 4c

5 7 × 6

6 7 × 3

7 7 × 4

8 7 × 5

9 24 take away 5

10 8 + 9 – 5

11 (7 + 3) × 2

12 \$8.93 = [ ] c

13 How many sixes in 18?

14 What is the product of 8 and 8?

15

I had \$50 but spent \$11. I have \$ [ ] left.

SET 2 Multiply tens then ones

Multiply the tens, then multiply the ones.
4 × 35 becomes
4 × 30 + 4 × 5 = 140

1 43 × 5

2 27 × 4

3 38 × 6

4 25 × 5

5 31 × 4

6 14 × 8

7 36 × 5

8 42 × 4

9 37 × 6

10 63 × 2

Working Mathematically

Round the 2-digit numbers to the nearest ten to give an approximate answer.

11 29 × 4 ≈

12 42 × 5 ≈

13 68 × 7 ≈

Space Compass points

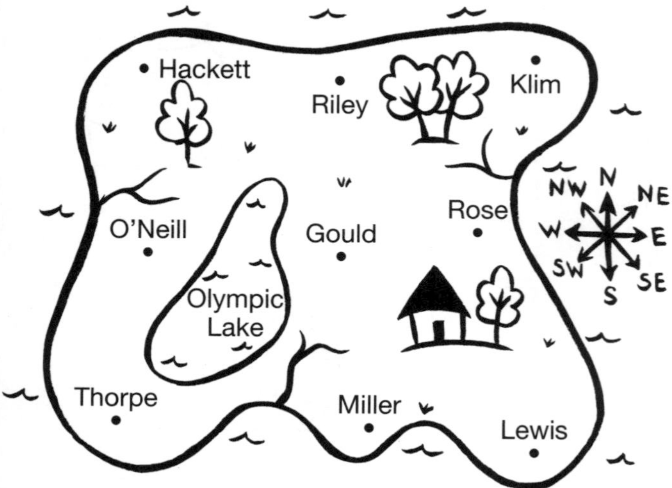

What direction are these destinations from Gould?

1 Riley ______

2 O'Neill ______

3 Thorpe ______

4 Hackett ______

5 What direction is Klim from Lewis?

______

6 What direction is Klim from Thorpe?

______

Number and Algebra

SET 3 Division

Write a multiplication fact for each division fact.

	Division		Multiplication
1	25 ÷ 5 = 5		5 × 5 = 25
2	32 ÷ 8 = ☐		☐ × ☐ = ☐
3	40 ÷ 10 = ☐		☐ × ☐ = ☐
4	28 ÷ 7 = ☐		☐ × ☐ = ☐
5	81 ÷ 9 = ☐		☐ × ☐ = ☐
6	42 ÷ 7 = ☐		☐ × ☐ = ☐
7	36 ÷ 6 = ☐		☐ × ☐ = ☐
8	54 ÷ 6 = ☐		☐ × ☐ = ☐

9 Forty-five crayons were shared among 6 children. How many did each person receive?

☐ remainder ☐

10 How many groups of 7 can be made from 23 people?

☐ remainder ☐

SET 4 Extension

1 (4 + 5) × 4

2 100 + 60 + 97

3 Estimate an answer to 399 + 148.

4 $\frac{1}{4}$ of 40 marbles

5 Add $4 to $10.95.

6 Half of $6.20

7 1323 + 501

8 Subtract $15 from $48.50.

9 70 + 70 + 70 + 70 + 70

10 Half of $8.60

11 1027, 1127, 1227, ☐, ☐

12 If 3 kg cost $6, how much would 7 kg cost?

13 How many legs on 2 dogs, 4 spiders and 2 children?

14 Cost of 3 books at $43 each

15 Write 2495 in words.

____

____

16 Write the largest number you can using 7, 0, 5, 9.

17 Is a glue stick shaped like a cone?

Measurement Perimeter

Measure the perimeters of these shapes.

1

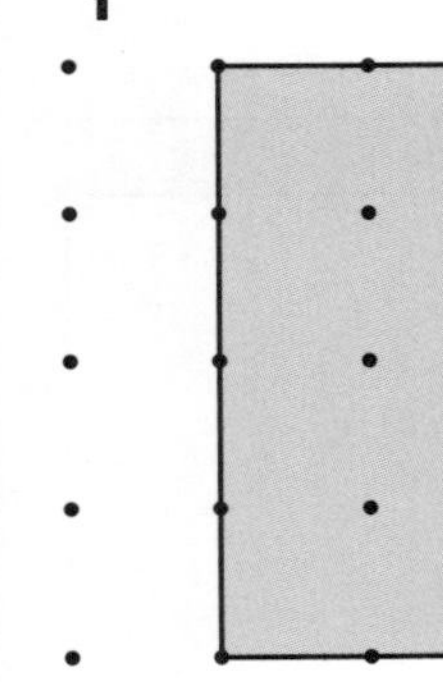

2

3

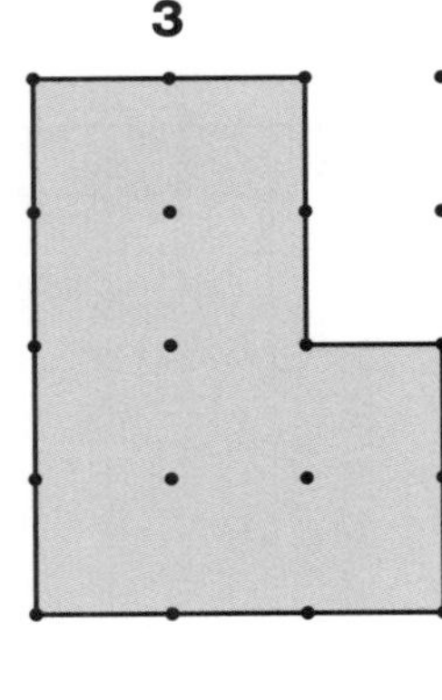

4

Number and Algebra

SET 1 Basic

1 9 + 5

2 5 × 7

3 (5 × 4) – 1

4 $7 + $7

5 4^2

6 5 + 4 + 3 + 2

7 Seven at 10c each

8 94 less 5

9 73 kg – 10 kg

10 What is the product of 8 and 9?

11 What is the difference between 35 and 6?

12 Halve 10 × 4.

13 Subtract 7 from 21.

14 962 = ☐ hundreds + ☐ tens + ☐ ones

15

The 3:12 train from the city was 10 minutes late. When did it arrive? ☐

SET 2 Addition

Add the numbers, you may need scrap paper.

1 Add 732 to 3835.

2 What is the sum of 1135 and 1700?

3 What is the cost of a $1635 stove and a $2317 fridge?

4 What is the total of 2430 and 3620?

5 2356 + 8 hundreds

Working Mathematically

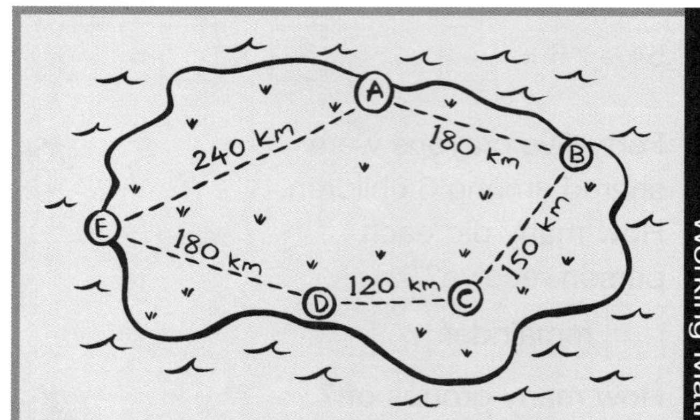

Use the map to decide if the statements about the trips are true or false.

	Trip		Trip	T or F
6	A to B to C	<	A to E to D	
7	E to D to C	>	D to C to B	
8	B to A to E	<	E to D to C	

Space Top, front and side views

Top view	Front view	Side view

Number and Algebra

SET 3 Number patterns

Follow the rules to complete the number patterns.

1

Multiply by 4							
2	4	6	8	10	12	14	16

2

Add 15							
1	2	3	4	5	6	7	8

3

Multiply by 4							
1	2	3	4	5	6	7	8

4

Multiply by 6							
3	4	5	6	7	8	9	10

Find the rule and complete the pattern.

5

Rule:							
1	2	3	4	5	6	7	8
5	6	7					

6

Rule:							
1	2	3	4	5	6	7	8
3	6	9					

SET 4 Extension

1 2 m and 37 cm = ☐ cm

2 How many millilitres in 4 L?

3 36 ÷ 6

4 (36 ÷ 4) + 5

5 4976, 4986, ☐, 5006

6 (3 × 10) + 9

7 How much are 3 cakes at $13 each?

8 What is $\frac{1}{4}$ of 120?

9 What number is halfway between 1350 and 1450?

10 $\frac{1}{5}$ of 25

11 How many angles on 7 hexagons?

12 There are 24 pegs on the clothes line.

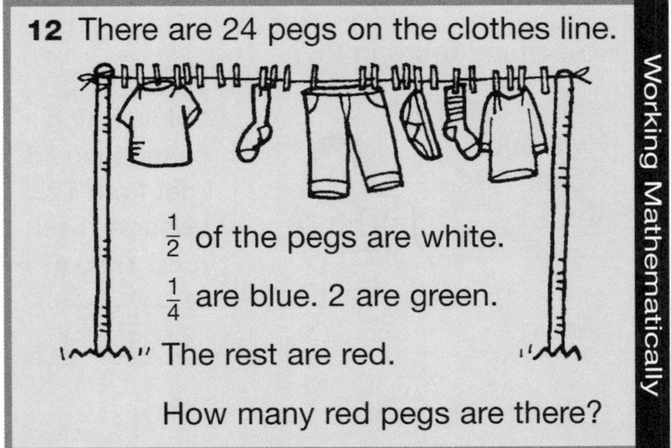

$\frac{1}{2}$ of the pegs are white.

$\frac{1}{4}$ are blue. 2 are green.

The rest are red.

How many red pegs are there?

Working Mathematically

Measurement Square centimetres

Draw two different shapes, each with an area of 15 square centimetres.

Number and Algebra

SET 1 Basic

1 9 + 7

2 8 + 5

3 7 × 6

4 7 × 4

5 7 × 8

6 7 × 7

7 68 – 6

8 52 – 6

9 8 + 8 + 8 – 4

10 7 + 10 – 3 + 4

11 $5.86 = ☐ c

12 How many 20c coins make $2.60?

13 What is the product of 8 and 7?

14 How many fours in 28?

15

How much change would I get from $2 if I bought 1 tea and 1 cake? ☐

SET 2 Rounding numbers

Round each number to the nearest 100 then the nearest 1000.

	Numbers	Nearest 100	Nearest 1000
1	976		
2	823		
3	1213		
4	1787		
5	2356		
6	7599		
7	4207		

Round each addition to the nearest 100 to make an estimate.

	Addition	Estimate
8	296 + 703 =	300 + 700 = 1000
9	237 + 759 =	
10	425 + 899 =	
11	901 + 989 =	
12	876 + 797 =	
13	1023 + 976 =	

Space Anti-clockwise turns

Draw the shape in its new position after making the anti-clockwise turns.

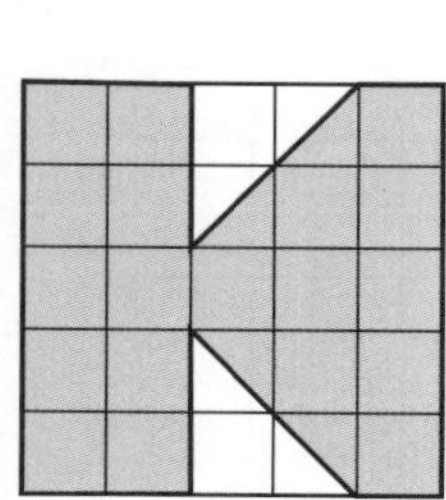

$\frac{1}{4}$ turn

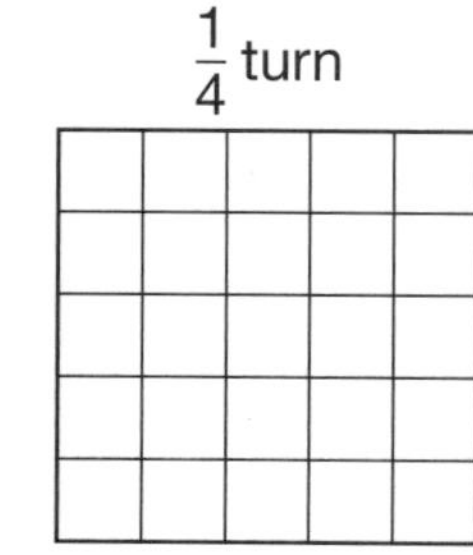

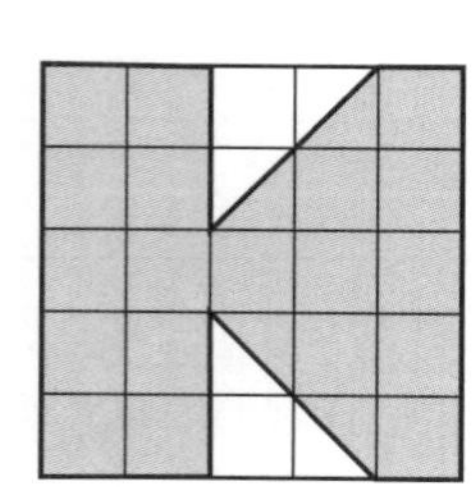

$\frac{1}{2}$ turn

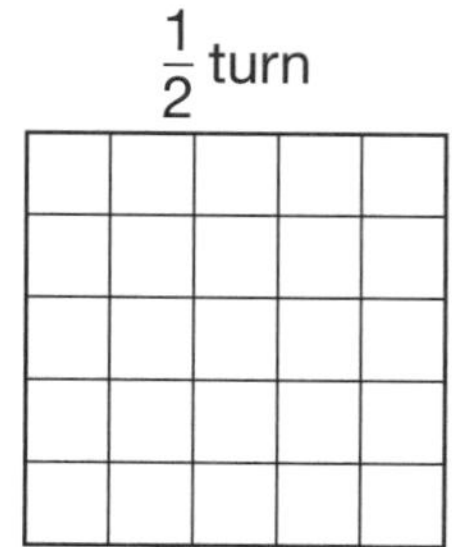

Number and Algebra

SET 3 Factors

Find four factors for each number.

	Number	Factors
1	20	
2	16	
3	12	
4	40	
5	32	
6	30	

7 Is 3 a factor of 12?

8 Is 8 a factor of 32?

9 Is 6 a factor of 27?

10 Is 6 a factor of 18?

Working Mathematically

11 Find all the factors for 24.

24 — 1, 2, ◯, ◯, ◯, ◯, ◯, ◯

SET 4 Extension

1 6 tens + 423

2 437 cm = ☐ m and ☐ cm

3 (36 ÷ 6) × 3

4 How many grams in $3\frac{1}{2}$ kg?

5 120 minutes = ☐ hours

6 What number represents thousands in 7523?

7 Share $36 among 6 people.

8 8846, 8746, 8646, ☐

9 How many hundreds in 4231?

10 How many legs on 9 beetles?

11 Estimate an answer to 137 + 249.

12 3796 + 203

13 If 10 apples cost $1, how much for 50 apples?

Working Mathematically

14 The perimeter of a rectangle is 120 cm. What are its dimensions if its length is twice its width?

length

width

length ________

width ________

Statistics and Probability Likelihood

Spinner A

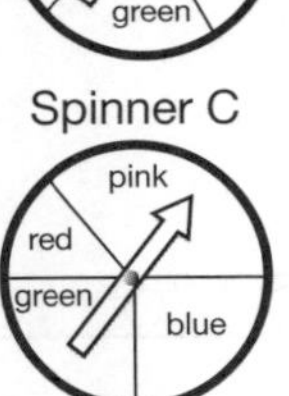

Spinner B

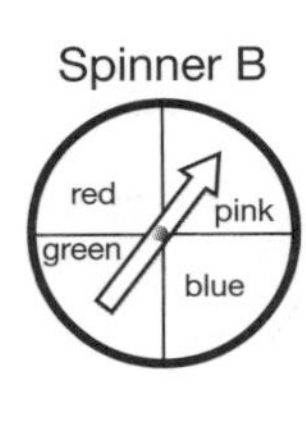

Spinner C

pink
red
green
blue

1 Which spinner is most likely to land on red? ☐

2 Which spinner is least likely to land on red? ☐

3 Which spinner is most likely to land on pink? ☐

4 Which spinner has equal chance of landing on any colour? ☐

5 Which spinner is least likely to land on pink? ☐

Number and Algebra

SET 1 Basic

1 6 + 2 + 8

2 9 + 9

3 16 – 6

4 (5 × 6) – 1

5 6c + 4c + 11c

6 25 take away 4

7 $9.27 = ______ c

8 How many sixes in 18?

9 Subtract 11 from 40.

10 7 tens × 6

11 Share $42 among 6 people.

12 Eight at $11 each

13 65 kg – 9 kg

14 Write 2060 in words.

____

____

SET 2 Subtraction

1 129 – 13

2 Subtract 26 from 597.

3 Subtract 9 from 374.

4 What is the difference between 186 and 8?

5 I had $347 and spent $42. How much money have I left?

6 What is $842 take away $54?

7

	HUND	TENS	ONES
	8	7	8
–	1	7	6

8

	HUND	TENS	ONES
	6	3	9
–	3	5	9

9

	HUND	TENS	ONES
	5	4	6
–	4	6	8

10

	HUND	TENS	ONES
	3	4	4
–	1	7	9

11 There are 249 children in the sports club. If 187 of them are girls, how many are boys?

Space Angles

Colour the larger angle in each pair.

1

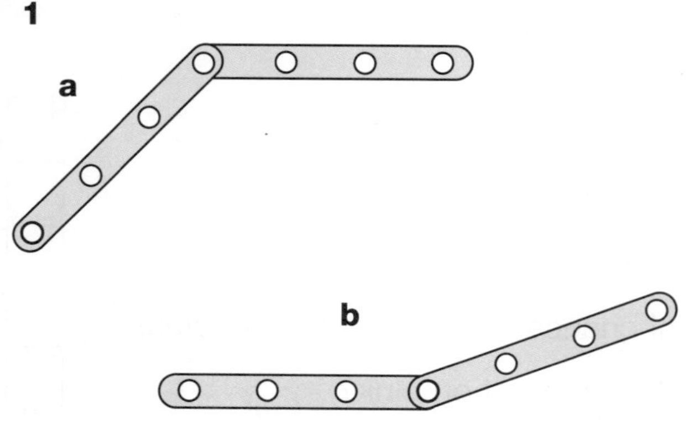

2

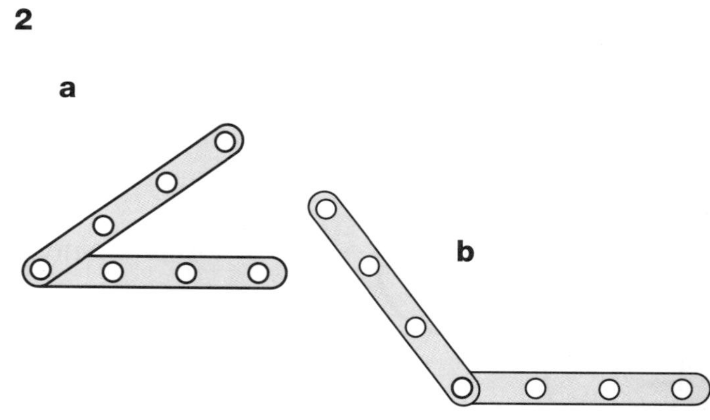

Number and Algebra

SET 3 Division

Complete the divisions and circle the correct bingo card.

1 $2\overline{)44}$ **2** $2\overline{)86}$ **3** $2\overline{)94}$

4 $3\overline{)36}$ **5** $3\overline{)93}$ **6** $3\overline{)42}$

7 $4\overline{)36}$ **8** $4\overline{)32}$ **9** $5\overline{)25}$

10 $5\overline{)35}$ **11** $6\overline{)78}$ **12** $6\overline{)90}$

13 $7\overline{)21}$ **14** $8\overline{)32}$

a

	8		22	
6		14		47

b

	5		43	
2		11		47

c

	7		15	
4		12		47

d

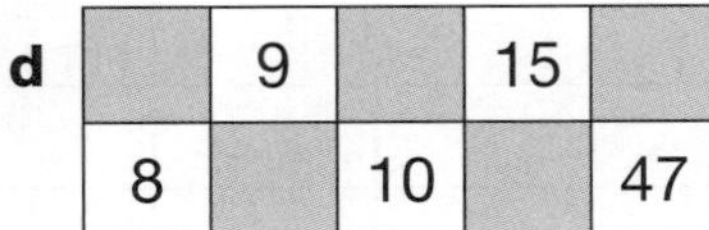

	9		15	
8		10		47

SET 4 Extension

1. (2 + 6) × 4
2. 1200 + 50 + 8
3. How many sixes in 48?
4. How many cents in $17.06?
5. Add $5 to $19.20.
6. Half of $6.10
7. (3 + 7) × 5
8. Subtract $9 from $19.95.
9. 20 + 20 + 20 + 20 + 20 + 20 + 20
10. $\frac{1}{4}$ of $4.68
11. How much is 3 kg at $2.50 per kg?
12. If 3 kg cost $33, how much would 7 kg cost?
13. 13 less than 9963
14. How many metres in 700 cm?
15. What is the perimeter of a square with 7 cm sides?

Working Mathematically

16 How old is Annabella if she is $\frac{1}{4}$ the age of Samira, who turned 28 yesterday?

Measurement Millimetres

Eight children were given a piece of paper to measure. Place each child's initial in the box that matches the measurement they made.

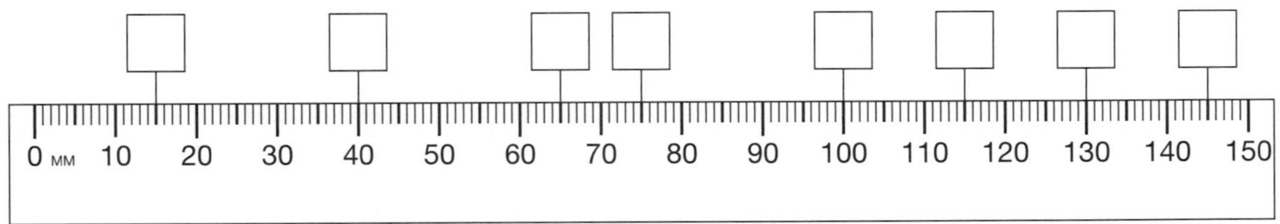

Lucy 40 mm	Jack 100 mm	Sam 145 mm	Emma 15 mm
Rose 115 mm	Bella 75 mm	Maya 65 mm	Hugo 130 mm

Number and Algebra

SET 1 Basic

1 18 – 8

2 18 – 10

3 12 + 6

4 18 + 18

5 7 × 2

6 6 × 8

7 4 × 9

8 7 × 3

9 6 + 7 + 6 + 2

10 23 take away 6

11 370c = $ ☐

12 How many fours in 24?

13 What is the sum of 19 and 6?

14 How many 20c coins in $1?

15 4057 = ☐ thousands + ☐ hundreds + ☐ tens + ☐ ones

SET 2 Compensation strategy

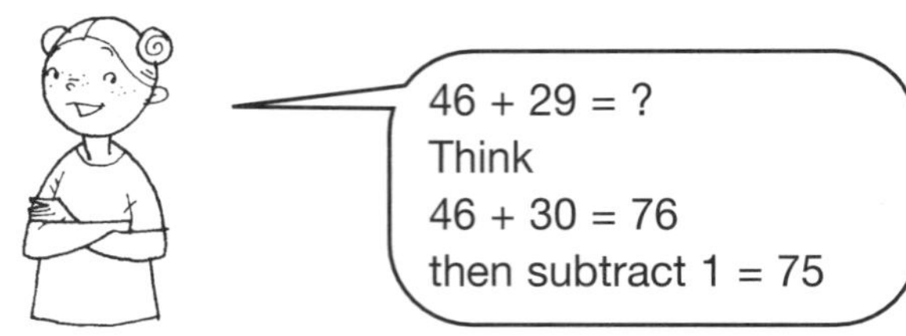

Round up the second number to help with these additions.

	Question	Becomes	Subtract	Answer
1	23 + 38	23 + 40 = 63	2	
2	43 + 49			
3	34 + 48			
4	27 + 59			
5	32 + 68			
6	117 + 39			

Round down the second number to help with these additions.

	Question	Becomes	Add	Answer
7	27 + 41	27 + 40 = 67	1	
8	44 + 32			
9	67 + 21			
10	48 + 52			
11	56 + 33			
12	137 + 42			

Space Acute, obtuse and right angles

Add another arm to create an angle that matches the label.

Acute angle	Right angle	Acute angle	Obtuse angle

Number and Algebra

SET 3 Hundredths

Shade the hundredths grids to represent the hundredths.

1 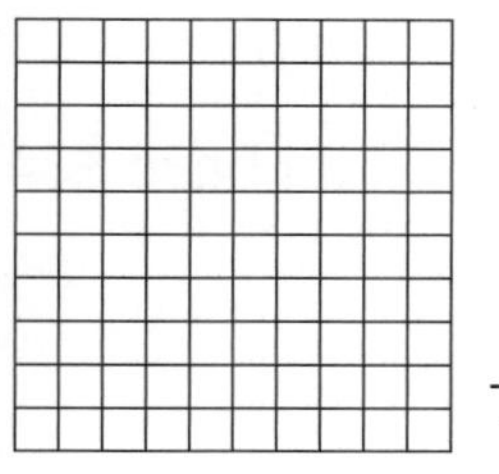 $\frac{25}{100}$

2 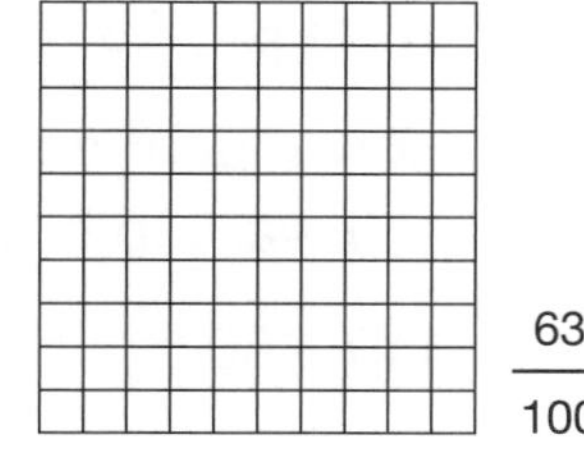 $\frac{63}{100}$

3 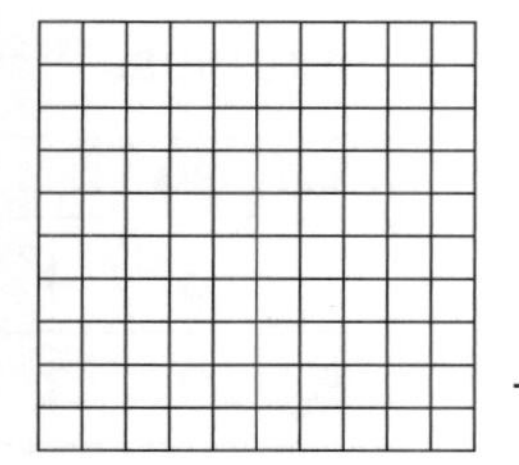 $\frac{55}{100}$

4 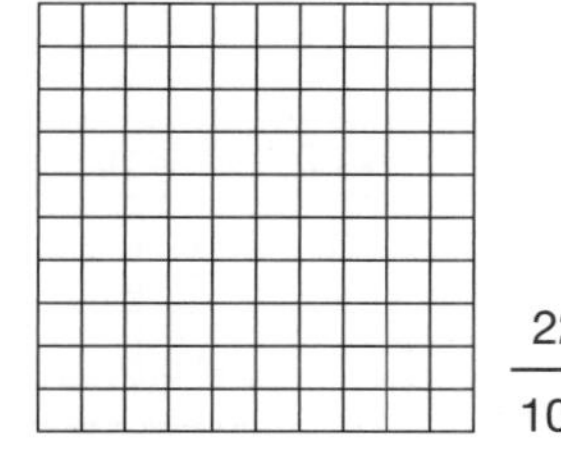 $\frac{22}{100}$

Trent measured 4 items using a metre ruler then recorded their measurements on a table.

Record each measurement as a fraction out of 100 cm (1 m).

5	33 centimetres	$\frac{\quad}{100}$
6	47 centimetres	—
7	69 centimetres	—
8	98 centimetres	—

SET 4 Extension

1 How many grams in 3 kg?

2 37, 137, 237, ☐

3 4 m and 5 cm = ☐ cm

4 $(30 \div 5) \times 6$

5 What is the value of 1 in 4107?

6 How many 25 cm lengths in 3 m?

7 What number is halfway between 126 and 130?

8 1236 + 603

9 $\frac{1}{5}$ of $80

10 How many days in winter?

11 $\frac{1}{2}$ kg meat at $11 per kilogram

12 How much is left from $5 if I spent $1.75?

13 Share 37 among 5.

Working Mathematically

14 Heather was supposed to cut a 180 cm piece of rope in half but she made a mistake and one piece was 20 cm longer than the other. How long are her pieces of rope?

0 cm 180 cm

Measurement Millilitres

Find the differences in capacity between these items.

1 A can of soft drink and a small milk carton ☐

2 A yoghurt tub and a medicine bottle ☐

3 A carton of cream and a Fizzo bottle ☐

4 A large milk carton and a carton of cream ☐

5 A soft drink can and a yoghurt tub ☐

6 A Fizzo bottle and a medicine bottle ☐

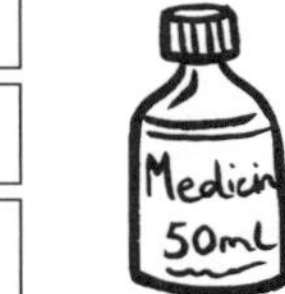

UNIT 22

Number and Algebra

SET 1 Basic

1 5×7

2 $20 - 5$

3 $9 \div 3$

4 7×5

5 $13 - 5$

6 $7 + 9 - 6$

7 What is the sum of 7 and 23?

8 What is the product of 8 and 3?

9 How many eights in 24?

10 95, 100, ☐, 110

11 A quarter of 12

12 $4.07 = ☐ c

13 What is the difference between 20 and 6?

14 3204 = ☐ thousands + ☐ hundreds + ☐ tens + ☐ ones

15

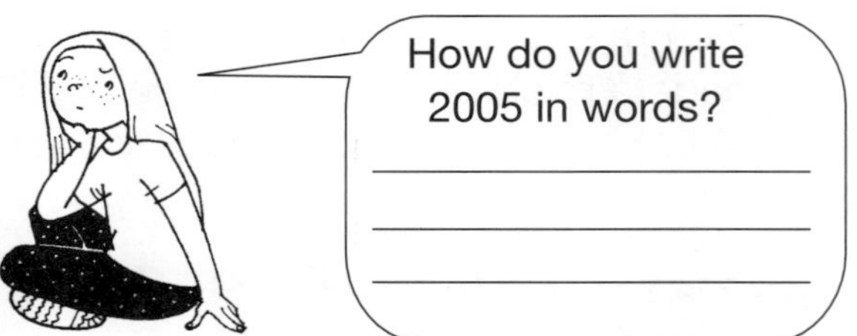

SET 2 4-digit subtraction

1 Melissa had $550 but spent $45. She has $ ☐ left.

2 What is $7.80 take away $2.80?

3 How much more do I need if I have $34 and want to buy a $73 gift?

4 440 – 60

5 500 g minus 240 g

6 Deduct 22 marbles from 77 marbles.

7

	HUND	TENS	ONES
	9	2	3
–		4	3

8

	HUND	TENS	ONES
	7	3	4
–	5	9	3

9

	HUND	TENS	ONES
	8	4	1
–	6	0	5

10

	HUND	TENS	ONES
	8	5	1
–	2	3	5

11

	HUND	TENS	ONES
	6	6	2
–	4	8	1

12

	HUND	TENS	ONES
	8	7	7
–	3	9	8

Space Tessellations

Replace the missing floor tiles. Colour them to make the pattern attractive.

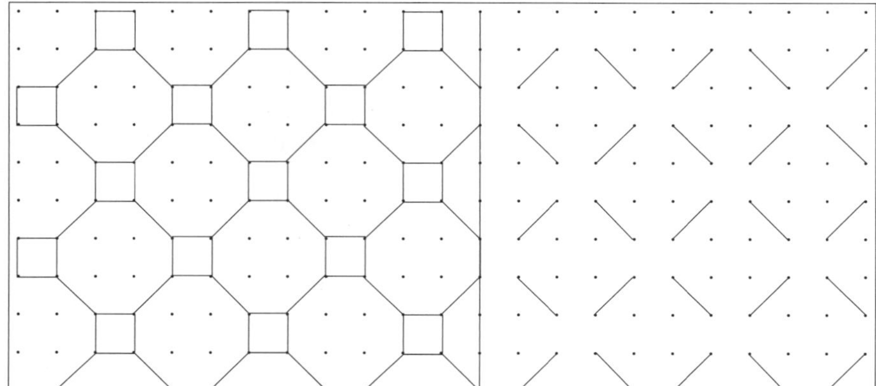

Number and Algebra

SET 3 Fractions of a collection

1 $\frac{1}{2}$ of 12 balls

2 $\frac{1}{4}$ of 12 balls

3 $\frac{1}{4}$ of 20 stars

4 $\frac{1}{5}$ of 20 stars

5 Joe had a bag of 16 marbles. If Jim had only $\frac{1}{4}$ of Joe's marbles, how many marbles did Jim have?

6 Sally baked 15 cakes but ate $\frac{1}{5}$ of them herself. How many did Sally eat?

7 Our school has 30 computers. If $\frac{1}{5}$ of them are broken, how many can still be used?

Working Mathematically

8 Bryan has saved $40. How much money has Jess saved if he has saved $\frac{1}{5}$ the amount that Bryan saved plus $10?

SET 4 Extension

1 How many 50 cm lengths in 5 m?

2 (18 – 5) + 6

3 6 kg at $5.50 per kg

4 Subtract $8 from $2479.

5 Share 56 marbles among 7 children.

6 How many sides have 8 octagons?

7 How many 250 g bags in a kilogram?

8 4435 + 402

9 What is the value of 5 in 7850?

10 Three positions after 429th

11 How much is left from $10 if I spent $1.65?

12 Half of 286

13 45 ÷ 3

14 How many tens in 489?

15 Write 4309 in words.

16 How much did Liam spend at the canteen if he bought the following?

1	SANDWICH	$1.50
1	SAUSAGE ROLL	$0.80
1	ORANGE	$0.45
1	FROZEN YOGHURT	$0.85

Measurement Using am and pm

Create a timetable using digital time for each part of your school day. Add am or pm after the time.

1 Wake up

2 Leave for school

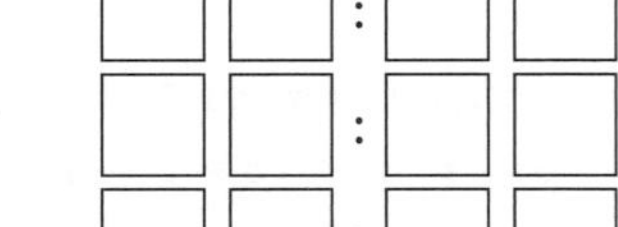

3 Have play lunch

4 Eat lunch

5 Finish school

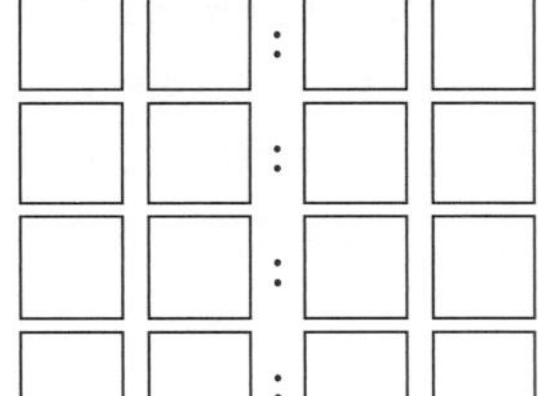

6 Eat dinner

7 Go to bed

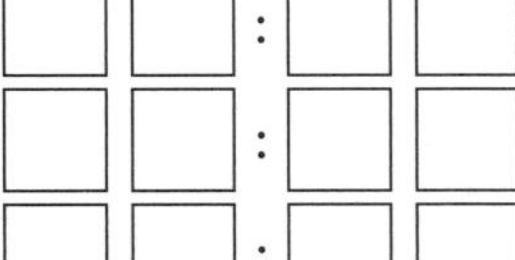

UNIT 23

Number and Algebra

SET 1 Basic

1 6 × 8

2 19 – 10

3 21 + 4

4 40 ÷ 4

5 8 × 5

6 31 – 12

7 21 + 6 + 7

8 What is the sum of 19 and 6?

9 What is the product of 8 and 10?

10 3, 6, ☐, 12, ☐, ☐, 21

11 A quarter of 20

12 Ten less than 50

13 Write one third of 24.

14 $2.02 = ☐ cents

15 6520 = ☐ thousands + ☐ hundreds + ☐ tens + ☐ ones

16 

How much do I have left from $50 if I spent $14.95?

$ ☐

SET 2 Multiplication

1 23×4

2 24×5

3 35×3

4 23×6

5 16×6

6 24×6

7 35×2

8 38×3

9 28×3

10 Billy bought 7 books that cost $25 each. How much did he spend? 

11 Lisa bought 5 skirts for $37 each. How much did she spend? 

Space Following directions

Where do you finish? ____________________

Number and Algebra

SET 3 Hundredths

Write a hundredth to describe each grid.

1 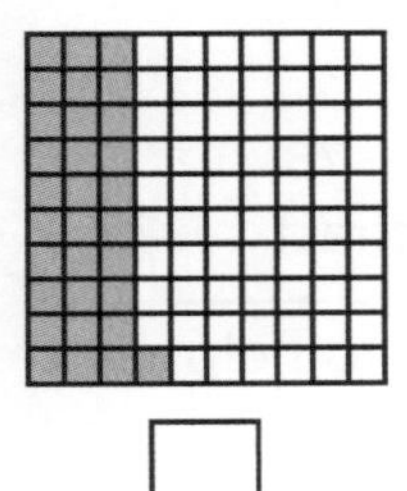

2 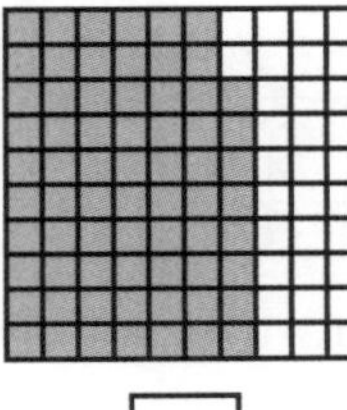

3 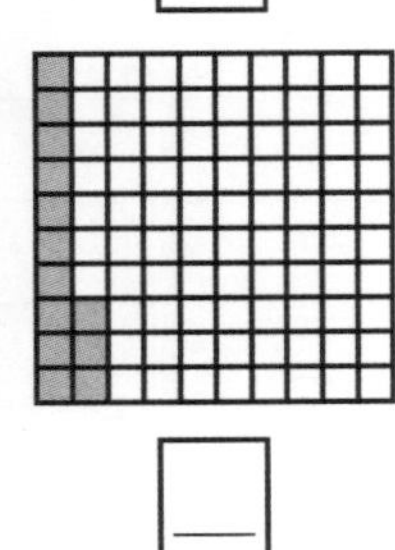

4 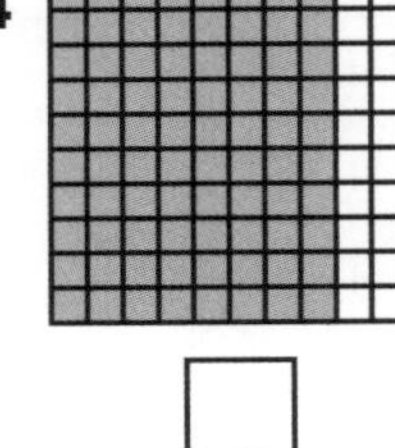

Write a decimal for each fraction.

5 $\frac{15}{100}$

6 $\frac{21}{100}$

7 $\frac{25}{100}$

8 $\frac{73}{100}$

9 $\frac{99}{100}$

SET 4 Extension

1 3127, 3147, 3167, ☐

2 Grams in 4 kilograms

3 247 cm = ☐ m and ☐ cm

4 What number represents thousands in 4306?

5 What is the perimeter of a pentagon with 7 cm sides?

6 How many faces has a rectangular prism?

7 If 4 items cost \$3, how much would 8 cost?

8 Share \$28 among 7 people.

9 $(42 \div 6) \times 2$

10 What is the change from \$5 if I spent \$1.85?

11 If 3 pears cost 60c, how much would 9 cost?

12 Write half-past five in digital time.

13 Is 40°C a hot day?

14 $\frac{1}{4}$ of 140

15 Write 497 in words.

____

____

16 Calculate the number of minutes between 1:55 am and 2:20 am.

17 Round 3743 to the nearest 100.

Statistics and Probability Picture graphs

Pocket money per week

Gina      

Max 

Zoe 

Lee 

Alex          

Number of dollars

1 How much does Gina earn per week? ______

2 Who earns the most money per week? ______

3 How much does Lee earn per week? ______

4 How much do Zoe and Max earn between them? ______

5 How much more does Alex earn compared to Max? ______

6 What is the total amount of money earned by the children? ______

UNIT 24

Number and Algebra

SET 1 Basic

1 6 × 4

2 20 − 12

3 20 + 18

4 50 ÷ 10

5 3 × 5

6 42 − 10

7 42 + 4 + 16

8 What is the sum of 25 and 11?

9 What is the product of 8 and 4?

10 90, 80, ☐, ☐, ☐, 40

11 A third of 18

12 Ten less than 48

13 49 kg + 11 kg =

14 3076 = ☐ thousands + ☐ hundreds + ☐ tens + ☐ ones

15

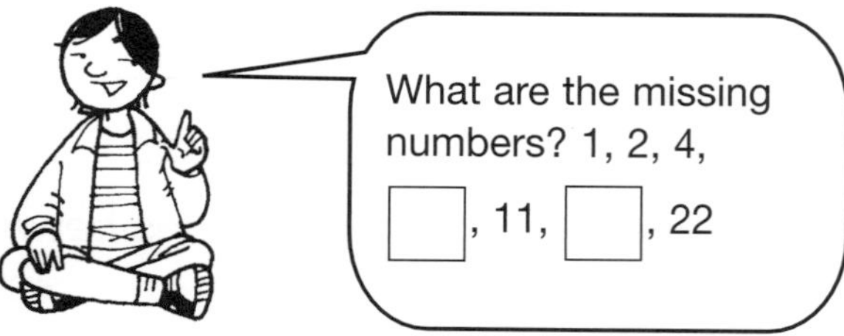

SET 2 Compensation strategy

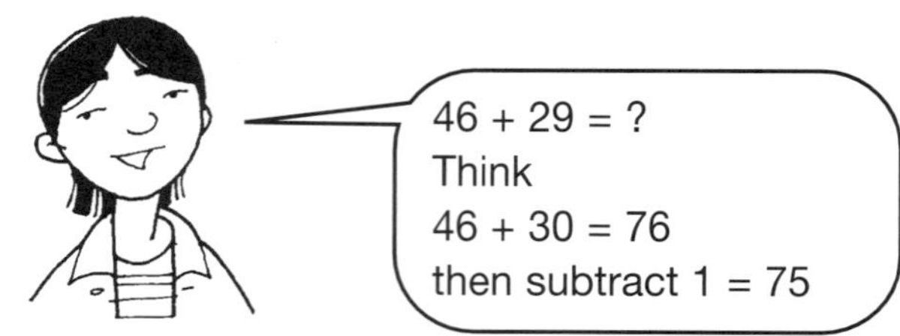

Round up the second number to help with these additions.

	Question	Becomes	Subtract	Answer
1	26 + 49	26 + 50 = 76	1	
2	37 + 58			
3	41 + 89			
4	43 + 47			
5	13 + 69			
6	114 + 47			

Round down the second number to help with these additions.

	Question	Becomes	Add	Answer
7	33 + 71	33 + 70 = 103	1	
8	25 + 42			
9	61 + 33			
10	76 + 22			
11	66 + 63			
12	123 + 51			

Statistics and Probability Survey/record data

Most popular snacks in Ms Johnson's class

Snack	Tally	Number
Cakes	𝍸 𝍸 |	
Biscuits	|||	
Pancakes	𝍸 𝍸	
Popcorn	𝍸 ||	

Construct a column graph from the data.

1 **Popular snacks**

Cakes												
Biscuits												
Pancakes												
Popcorn												

Number

2 Which was the most popular food?

3 How many children are in Ms Johnson's class?

Number and Algebra

SET 3 Division

Crack the code.

A	D	E	F	H	L	N	O	R	S	U
8	5	1	9	2	6	11	3	4	10	7

1 24 ÷ 6 **2** 63 ÷ 9

3 40 ÷ 8 **4** 21 ÷ 7

5 30 ÷ 5 **6** 81 ÷ 9

7 20 ÷ 10 **8** 32 ÷ 4

9 90 ÷ 9 **10** 40 ÷ 5

11 32 ÷ 8 **12** 7 ÷ 7

13 25 ÷ 5 **14** 55 ÷ 5

15 27 ÷ 9 **16** 20 ÷ 2

17 10 ÷ 10

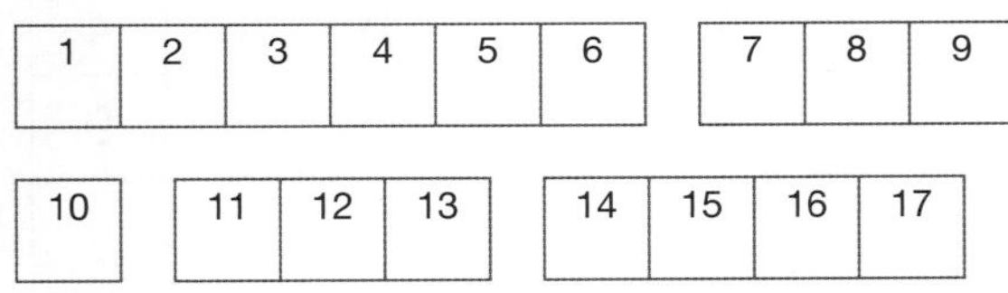

18 Sixty-four children need to be transported by a minibus that carries 8 passengers. How many trips will need to be made to transport all the children?

SET 4 Extension

1 How many 75 cm lengths in 3 m?

2 What is the difference between 187 and 45?

3 7 decades = ☐ years

4 (3 × 4) + 7

5 How many corners has a cube?

6 Does a decagon have 10 sides?

7 7 pears at 85c each

8 Write all the factors for 18.

9 $\frac{1}{4}$ of $96

10 1375 mL = ☐ L + ☐ mL

11 Is 12 a multiple of 6?

12 What is the value of 0 in 7603?

13 What is the smallest number of coins needed to make $4.65?

Working Mathematically

14 If a car travels at 50 km/h for 4 hours, how far will it travel?

15 How far will a train travel in 4 hours at 100 km/h?

Measurement Grams

Record these measurements in abbreviated form.

1 1 kilogram and 600 grams = 1 kg 600 g

2 2 kilograms and 260 grams = ______________

3 4 kilograms and 950 grams = ______________

4 5 kilograms and 20 grams = ______________

5 3 kilograms and 35 grams = ______________

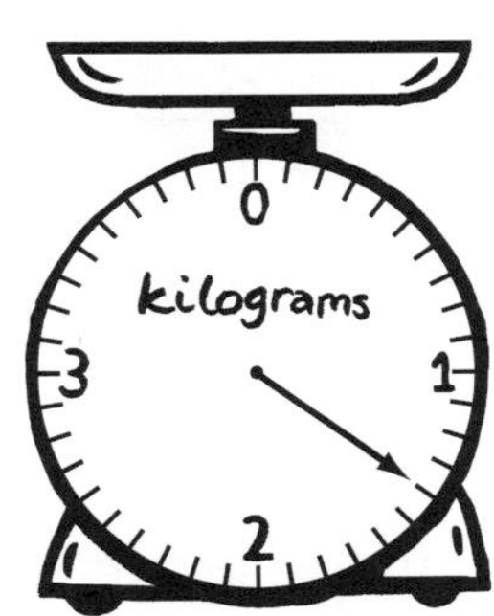

UNIT 25

Number and Algebra

SET 1 Basic

1 7×7

2 17 – 6

3 17 + 6

4 14 – 7

5 8×7

6 6 + 6 + 6 + 6 – 7

7 9×7

8 What is the product of 7 and 10?

9 What is the difference between 49 and 20?

10 What is the product of 7 and 4?

11 How many 5c coins in 65c?

12 A quarter of 16

13 $2.59 = ☐ c

14 320 = ☐ thousands + ☐ hundreds + ☐ tens + ☐ ones

15

What odd number is between 40 and 46 and greater than 43?

☐

SET 2 Addition

1 300 + 400 + 200

2 300 + 45 + 43

3 Add 325, 50 and 120.

4 What is the total of 121, 234 and 423?

5

	HUND	TENS	ONES
	3	4	7
	5	2	0
+	1	2	8

6

	HUND	TENS	ONES
	3	6	7
	2	0	4
+	3	6	2

7

	HUND	TENS	ONES
		2	7
	5	8	1
	2	4	4
+		2	9

8

	HUND	TENS	ONES
	3	5	8
		6	4
	6	3	4
+			9

9 Peter earned $328, $256 and $175. How much did he earn in total?

Peter earned $ ☐.

Space Pentagons and octagons

1 Colour the octagons and put a P on the pentagons.

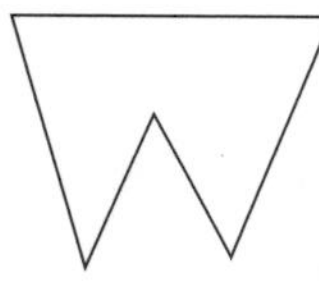

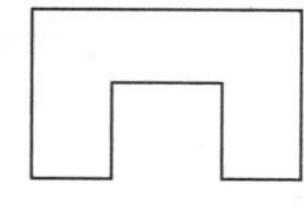

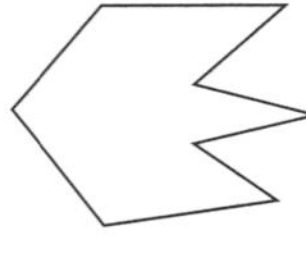

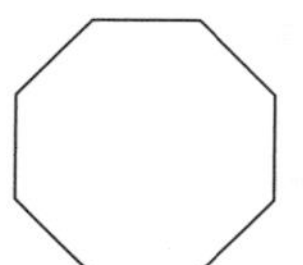

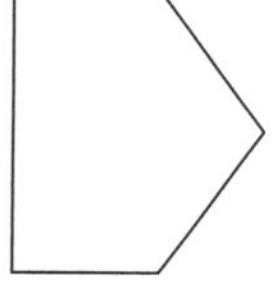

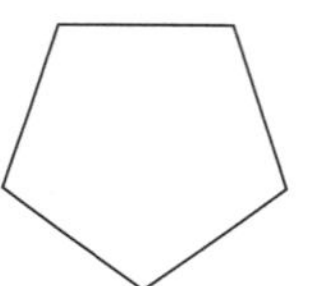

2 Colour the 3D objects that have a pentagon as a face.

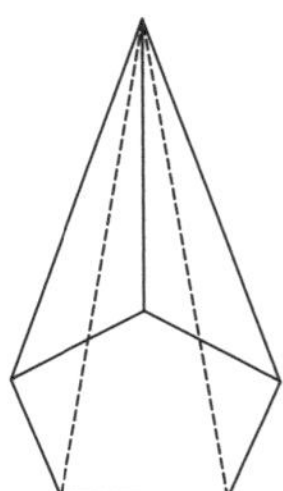

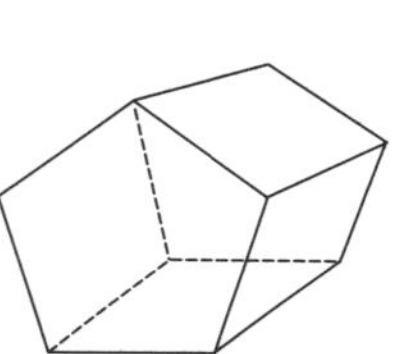

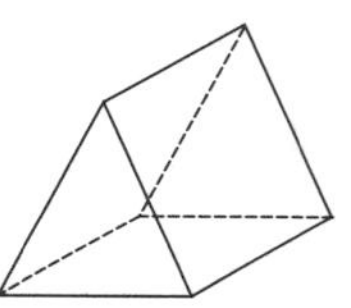

Number and Algebra

SET 3 Decimal place value

Colour the largest fraction or decimal.

1	0.71	$\frac{80}{100}$	0.52
2	0.17	1.0	0.71
3	0.19	0.11	0.21
4	0.09	0.43	0.25
5	$\frac{33}{100}$	0.3	0.29
6	4.01	1.04	0.41

Give the place value of each bold digit.

7 7.2**1** ______

8 **1**.27 ______

9 7.**2**5 ______

10 3.0**7** ______

Order the decimals from smallest to largest.

11 0.35, 0.29, 0.09 ______

12 0.33, 0.39, 0.29 ______

13 0.87, 1.07, 7.01 ______

14 9.85, 8.59, 9.58 ______

SET 4 Extension

1 Estimate an answer to 137 + 203.

2 $3\frac{1}{4}$ kg = ☐ g

3 How many 250 g bags in 2 kg?

4 Half of a kilometre

5 Share 40 among 6.

6 How many quarters in $1\frac{3}{4}$?

7 How many minutes in half an hour?

8 $5^2 + 3$

9 $(5 \times 8) + 4$

10 4302 + 583

11 30 + 5 + 108

12 What is the product of 9 and 6?

13 Arrange 5, 6, 4, 9 to make the largest number possible.

Working Mathematically

14 This model was built with 7 cubes. If Lea numbered every square on the model's surfaces, how many numbers did she use?

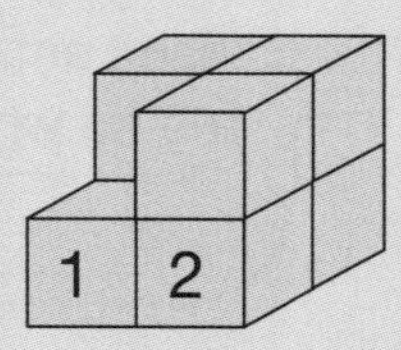

Measurement Square metre

Tick the box which best describes the area.

	Less than 1 m^2	About 1 m^2	More than 1 m^2
TV screen			
Notice board			
Skateboard			
Door			
Photo frame			
Bedroom floor			

UNIT 26

Number and Algebra

SET 1 Basic

1 7 × 2

2 30 – 15

3 36 + 11

4 60 ÷ 6

5 8 × 8

6 58 – 10

7 56 + 14

8 What is the sum of 37 and 13?

9 8, 16, ☐, 32

10 What is a third of 24?

11 Ten less than 138

12 A quarter of 80

13 $3.79 = ☐ c

14 Write 176 in words.

____

____

15 My sister is 5 years older than I am. How old will she be when I am 19? ☐

SET 2 Decimals

Write the decimal that is modelled by the shaded section of each hundredths grid.

1 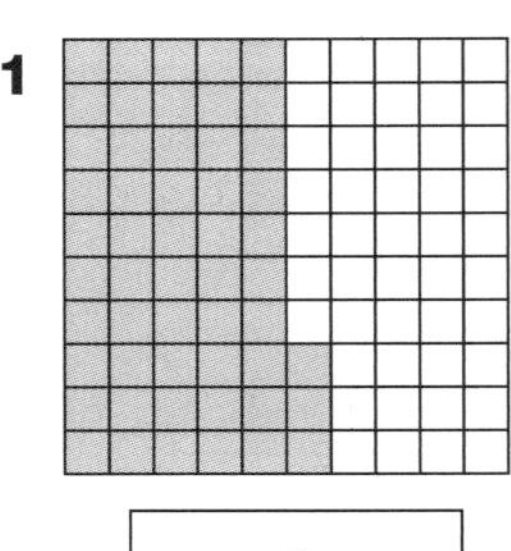

☐ . ☐

2 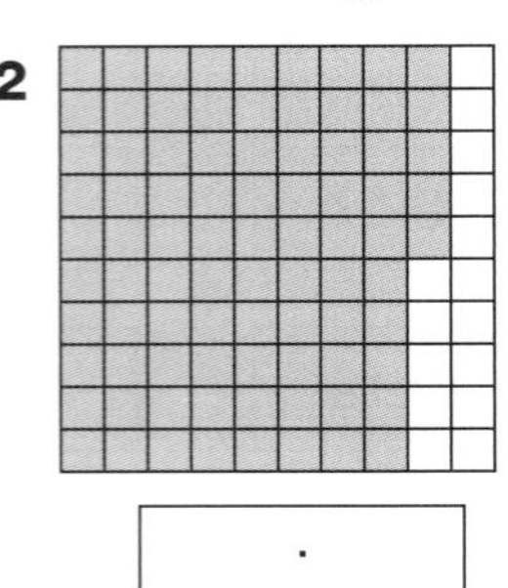

☐ . ☐

3 Write $\frac{21}{100}$ as a decimal.

4 Write $\frac{32}{100}$ as a decimal.

5 Write $\frac{54}{100}$ as a decimal.

6 Write $\frac{68}{100}$ as a decimal.

7 Which is larger, 0.46 or 0.64?

8 Which is larger, 0.57 or 0.75?

9 Which is larger, 0.69 or 0.96?

10 Order these decimals from smallest to largest: 0.83, 0.91, 0.85, 0.72.

Space Tessellations

Continue this tessellating pattern.

Number and Algebra

SET 3 Finding unknown quantities

Supply the missing numbers so that both sides of the balance beam are equal.

1. 25 + 15 | 17 +
2. 30 + 12 | 58 –
3. 68 – 24 | + 13
4. 76 – 20 | – 25
5. 69 + 21 | + 45
6. 72 – 55 | 86 –
7. 66 – 49 | + 0

SET 4 Extension

1. $546 – $12
2. $4\frac{1}{4}$ kg = ☐ g
3. List the factors of 15.
4. $\frac{1}{10}$ of a kilogram = ☐ g
5. Share 50 among 6 people.
6. How many tenths in a whole?
7. How many minutes in $\frac{3}{4}$ of an hour?
8. Round 1969 to the nearest 100.
9. $6^2 + 3$
10. List the factors of 20.
11. If 9 pencils cost 81c, how much would 20 pencils cost?
12. What is the product of 9 and 4?
13. Arrange 6, 1, 7, 2 to make the smallest number possible.
14. If 5 rulers cost 95c, how much would 20 cost?

Working Mathematically

15. Lee is 5 years older than Jack, who is 15 years younger than Nick, who is 2 years older than Con.

    How old is Lee if Con is 45? ________

Measurement Cubic centimetres

Record the volume of each object in cubic centimetres.

1. 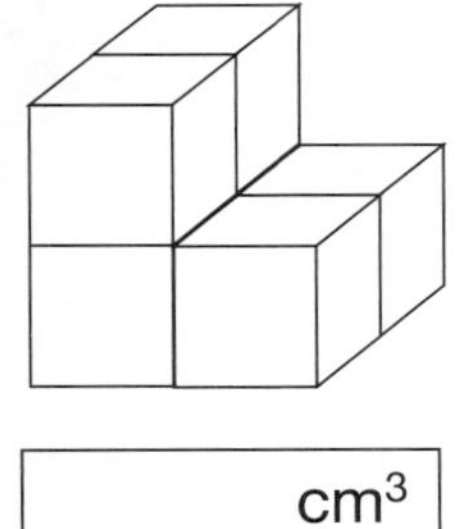

   ☐ cm^3

2. 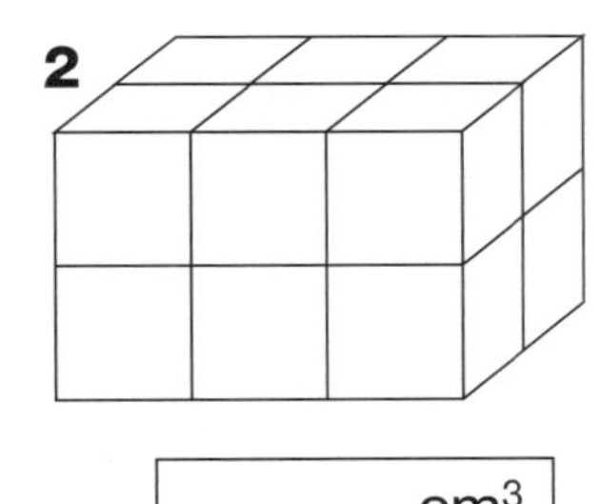

   ☐ cm^3

3. 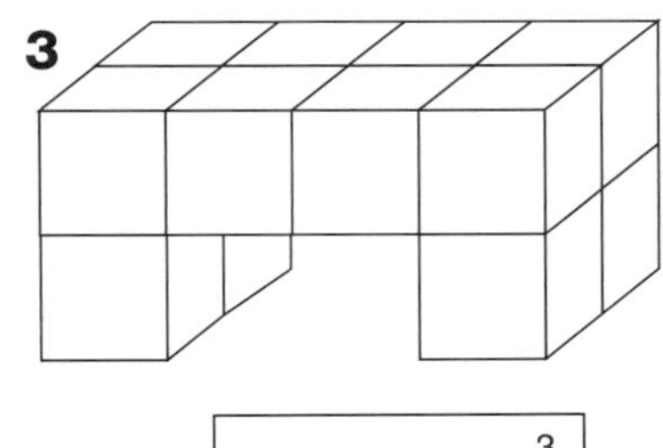

   ☐ cm^3

4. 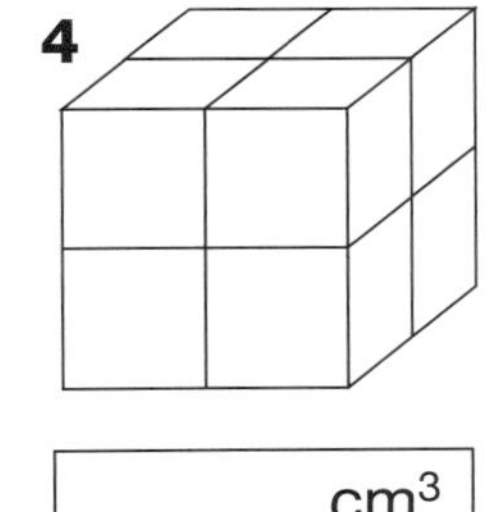

   ☐ cm^3

Number and Algebra

SET 1 Basic

1 50 – 25

2 10 – 8

3 28 – 21

4 24 + 9

5 18 + 6

6 9 × 3

7 9 × 5

8 9 × 4

9 What is the product of 7 and 9?

10 What is the sum of 17 and 13?

11 A third of 30

12 309c = $ ☐

13 What is the difference between 40 and 25?

14 27 = ☐ thousands + ☐ hundreds + ☐ tens + ☐ ones

15

How old am I if I'm 16 years younger than Aunt Kylie, who is 25? ☐

SET 2 Multiplication

	×	0	5	3	6	4	7	9
1	3							
2	5							
3	7							
4	9							
5	8							

6

HUND	TENS	ONES
	5	6
×		6

7

HUND	TENS	ONES
	2	4
×		3

8

HUND	TENS	ONES
	7	5
×		4

9 The principal, Mr Horvath, used 5 sheets of stickers on award day. How many stickers were used if there were 42 on each sheet?

10 Jonas saved $38 every month for 6 months. How much money did he save?

Space Grid references

1 Colour the space C4 blue.

2 Colour the space B5 red.

3 Colour the space D3 green.

4 Put an X in G7.

5 Draw a circle in H1.

6 Put the letter A in D8.

7 Put the letter E in E5.

8 Draw a triangle in A7.

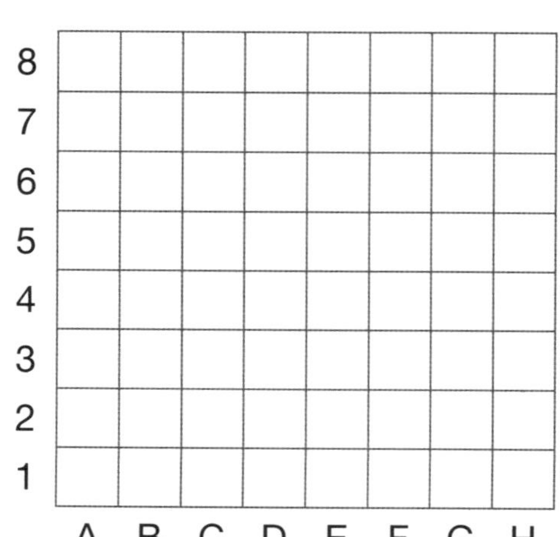

Number and Algebra

SET 3 Division

Use the halve and halve again strategy to solve these divisions.

1 36 ÷ 4

2 48 ÷ 4

3 64 ÷ 4

4 Share 84 stamps among 4 people.

Use the halve, halve and halve again strategy to solve these divisions.

5 Share $72 among 8 people.

6 $96 ÷ 8

7 $112 ÷ 8

8 $240 ÷ 8

9 $208 ÷ 8

10 $256 ÷ 8

Do these divisions.

11 $5\overline{)70}$

12 $4\overline{)72}$

13 $6\overline{)82}$

14 $7\overline{)95}$

SET 4 Extension

1 ($\frac{1}{2}$ of 24) × 0

2 How many corners has a rectangular prism?

3 $5^2 + 5$

4 3000 mL = ☐ L

5 Is 42 a multiple of 6?

6 21 hundredths = 0.☐

7 What is the value of 9 in 9327?

8 How much are 3 pieces of cake at 35c each?

9 How much is $2\frac{1}{2}$ kg of flour at $1.20 per kg?

10 Write all the factors for 30.

11 How many millilitres in $5\frac{1}{2}$ L?

12 If 8 cost $96, how much would 5 cost?

13 A boy ran for 3 hours at an average speed of 12 km/h. How far did he run?

Working Mathematically

14 Gillian is trying to solve 56 ÷ 8 on her calculator but the 8 button is missing.

Place numbers in the boxes so that her problem can be solved.

☐ ÷ ☐ ÷ ☐ = 7

Statistics and Probability Chance

Taylor has a bag of 11 marbles coloured red, blue, pink and green.

1 Which colour is most likely to be drawn out of the bag? ______________

2 Which colours are least likely to be drawn out of the bag? ______________

3 If 3 red marbles were drawn out of the bag and thrown away, what colour would then be most likely to be drawn out of the bag? ______________

Number and Algebra

SET 1 Basic

1 38 – 24

2 What is the product of 6 and 4?

3 20 ÷ 4

4 6, 12, ☐, 24, ☐

5 Ten less than 254

6 45 ÷ 5

7 What is the difference between 38 and 19?

8 A quarter of 20

9 $3.65 = ☐ c

10 42 ÷ 6

11 A third of 21

12 What is the sum of 8, 9 and 10?

13 Eleven less than 44

14 659c = $ ☐

15 Write four thousand, three hundred and twenty-one as a numeral.

16 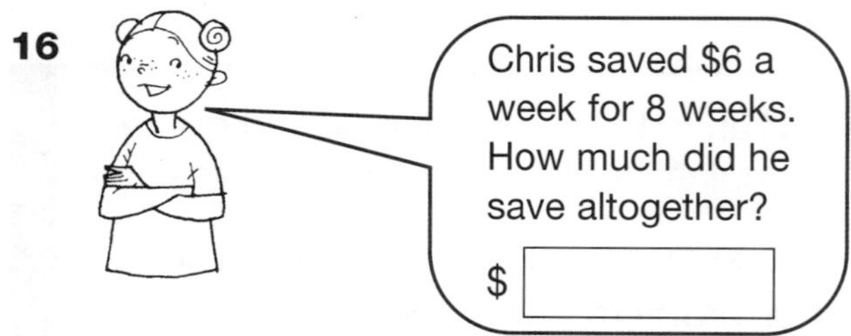

SET 2 4-digit addition

1 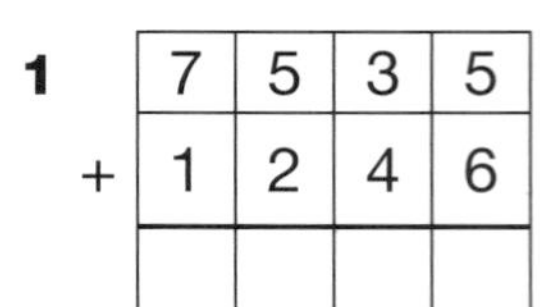

	7	5	3	5
+	1	2	4	6

2 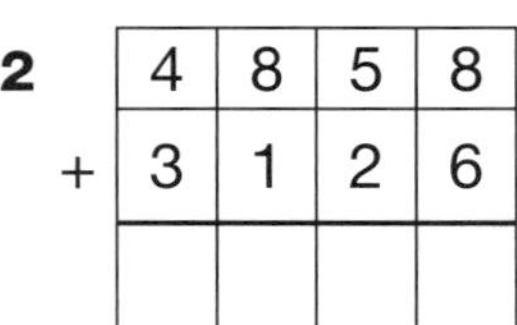

	4	8	5	8
+	3	1	2	6

3

	2	6	7	9
+	0	3	1	6

4 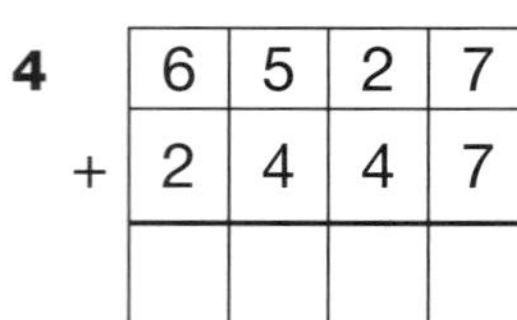

	6	5	2	7
+	2	4	4	7

5 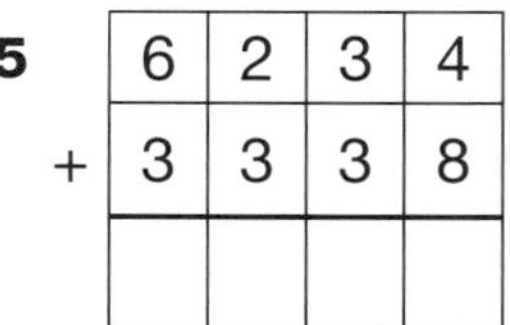

	6	2	3	4
+	3	3	3	8

6 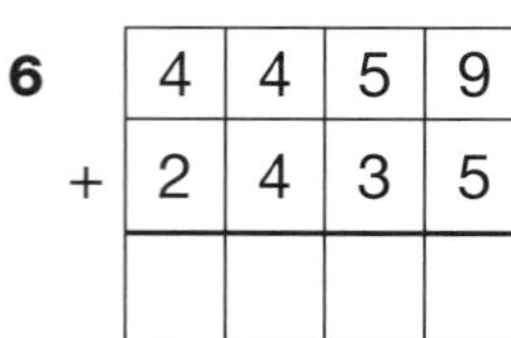

	4	4	5	9
+	2	4	3	5

Supply the missing numbers in these additions.

7

		1	2	
+	3	8		3
	7	9	8	8

8 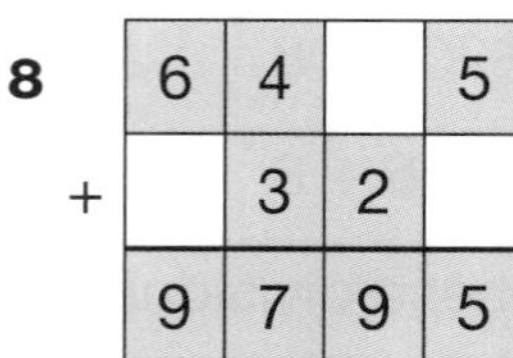

	6	4		5
+		3	2	
	9	7	9	5

Statistics and Probability Representing data

1 Ms Toms' class conducted a survey to find out which cars were driven by the teachers. Use the table below to construct a column graph.

Make	Ford	Toyota	Tesla	Nissan	Mazda	Other
Number of cars	𝍸	||||	𝍸	||	|	||||

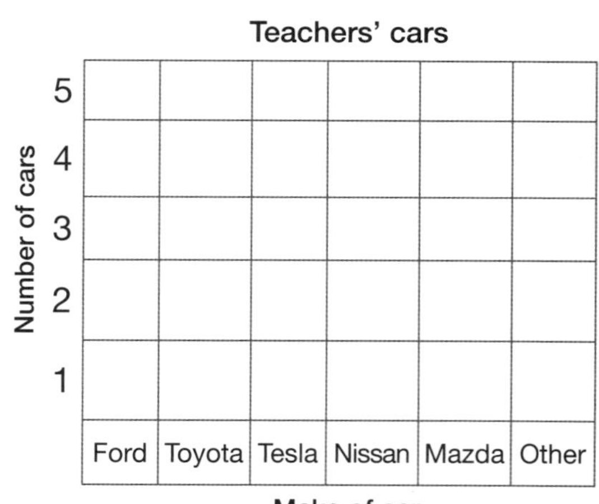

2 What were the two most popular makes of cars driven by the teachers?

Number and Algebra

SET 3 Commutative laws

Complete the sets of additions.

1	3 + 7 =	7 + 3 =
2	9 + 8 =	8 + 9 =
3	20 + 70 =	70 + 20 =
4	80 + 70 =	70 + 80 =
5	126 + 6 =	6 + 126 =
6	250 + 26 =	26 + 250 =

7 Did the order of adding the numbers change the answer?

Complete the sets of multiplications.

8	$6 \times 5 =$	$5 \times 6 =$
9	$7 \times 4 =$	$4 \times 7 =$
10	$8 \times 3 =$	$3 \times 8 =$
11	$8 \times 6 =$	$6 \times 8 =$
12	$7 \times 5 =$	$5 \times 7 =$
13	$8 \times 4 =$	$4 \times 8 =$

14 Did the order of multiplying change the answer?

SET 4 Extension

1 $6.25 + $4.50

2 6 kg at $4.20 per kg

3 How many tens in 568?

4 Round 1944 to the nearest 10.

5 $(54 \div 6) + 8$

6 1, 4, 9, ☐, 25, ☐, 49

7 What is the change from $5 if I spent $2.45?

8 How many thousands in 85 901?

9 Share 48 among 5.

10 What is the value of 6 in 16 250?

11 Three watches at $99 each

12 8 decades = ☐ years

13 Round to the nearest 10 to find an estimate for 687 – 253.

14 How many 60 cm lengths can be cut from 3 m?

Working Mathematically

15 Mr Klim spent $36 on a book and a calculator. How much did the book cost if the calculator cost twice as much?

Book $ ☐

Measurement Kilograms and grams

Label the mass shown on each scale.

1

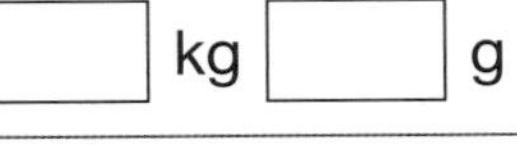

2

☐ kg ☐ g

3

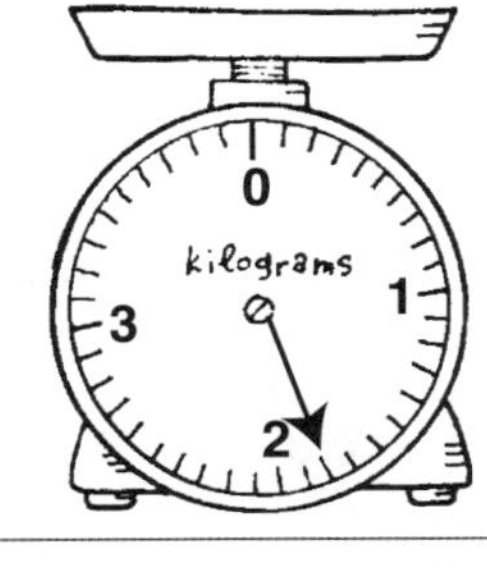

☐ kg ☐ g

4

☐ kg ☐ g

Number and Algebra

SET 1 Basic

1 $4 \times 10c$

2 5c + 15c

3 $3 \times \$2$

4 $\$20 \div 2$

5 \$18 + \$12

6 \$4.50 – \$2.00

7 $\$5 \times 6$

8 $\$21 \div 3$

9 What is the product of 6 and 11?

10 What is the sum of 15 and 14?

11 What is the difference between 35 and 42?

12 One-quarter of \$36

13 Share \$24 among 3 people.

14 Write \$9.56 in words.

____

____

15 We ordered two pizzas. How much pizza is left if we ate one half of each pizza?

SET 2 Square numbers

1 Complete the square number table.

2^2	3^2	4^2	5^2	6^2	7^2	8^2
4	9					

2 What number has the factors 9 and 5?

3 Triple 9.

4 What number is 6 times larger than 8?

5 5 squared

6 What number is 4 times larger than 10?

7 How many sevens are needed to make 49?

8

	HUND	TENS	ONES
		3	5
×			4

9

	HUND	TENS	ONES
		4	5
×			5

Working Mathematically

10 Jian's sister spilt ink on his homework. Write in the missing numbers to help Jian.

	HUND	TENS	ONES
		$^{2}5$	
×			6
		2	4

Number and Algebra Rounding to 5 cents

Round each price to the nearest 5 cents. Record the rounded price on the price tag below each item.

↓	↑
1c	3c
2c	4c
6c	8c
7c	9c

\$1.84

\$2.46

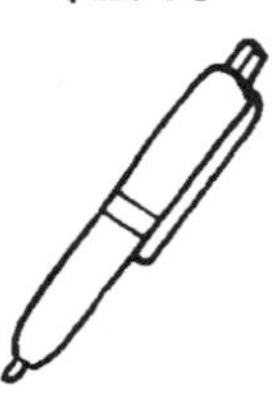

\$1.78

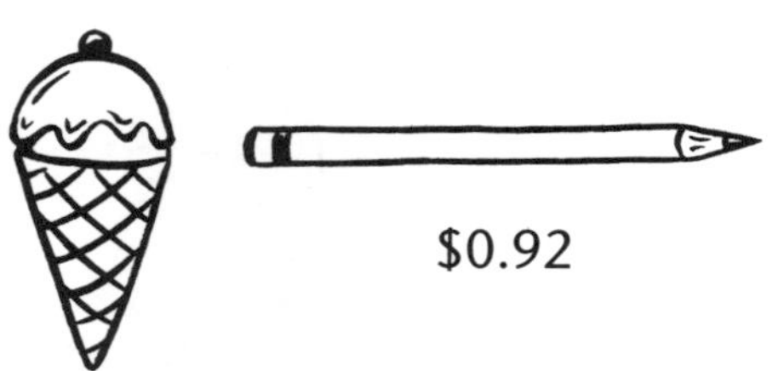

\$0.92

\$1.13

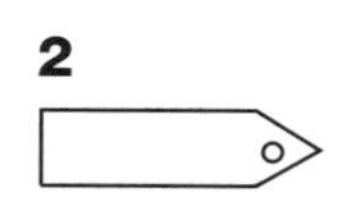

1

2

3

4

5

Number and Algebra

SET 3 Division

Find the winning bingo card by completing the number sentences.

A

	8		10 r 1		6	
5 r 4		8 r 1		9		3

B

	8		3		5 r 4	
6		9 r 2		10 r 1		2

C

	9 r 2		10 r 1		8	
3		9		6		15

1 $12 \div 4 =$ ☐ **2** $44 \div 8 =$ ☐

3 $33 \div 3 =$ ☐ **4** $26 \div 2 =$ ☐

5 $40 \div 5 =$ ☐ **6** $36 \div 6 =$ ☐

7 $25 \div 3 =$ ☐ **8** $27 \div 3 =$ ☐

9 $47 \div 5 =$ ☐ **10** $31 \div 3 =$ ☐

Solve these divisions.

11 $4\overline{)90}$ **12** $6\overline{)92}$ **13** $6\overline{)79}$

14 $2\overline{)77}$ **15** $4\overline{)65}$ **16** $3\overline{)52}$

SET 4 Extension

1 $(3 \times 0) \times 5$

2 37 hundredths = 0. ☐

3 $3\frac{1}{2}$ m = ☐ cm

4 6 pencils at 45c each

5 How many faces has a square pyramid?

6 Write all the factors of 21.

7 (☐ $\times$ 6) + 1 = 37

8 How much is 3 kg of sugar at $4.30 per kg?

9 Is 42 a multiple of 7?

10 What fraction of 12 is 6?

11 What is the value of 7 in 3075?

12 Write 3 tenths as a decimal.

13 1 tenth of 100

14 How many 10c coins in $12.50?

15 Perimeter of an octagon with 13 cm sides

16 Complete the magic square.

	8	18
		20
		10

17 If 4 kg of meat costs $20.00, how much is 7 kg?

Statistics and probability Data

Four of the most popular ways that people get information are through Facebook, Instagram, X and TikTok. The pie chart shows the results of an online survey of 96 people.

Complete the table below based on the information presented in the pie chart about the 96 people in the survey.

Source	Facebook	Instagram	X	TikTok
Number of people				

Popular sources of information

Number and Algebra

SET 1 Basic

1 5 × 11c

2 6c + 6c + 5c

3 \$2 × 3

4 \$18 ÷ 6

5 \$72 + \$19

6 \$4.50 – \$3

7 \$15 × 2

8 \$21 ÷ 7

9 What is the product of 8 and 3?

10 What is the sum of 13, 18 and 19?

11 What is the difference between 49 and 100?

12 What is one-fifth of \$10?

13 Share \$72 among 9 people.

14 Write \$16.58 in words.

____

____

SET 2 4-digit addition

1
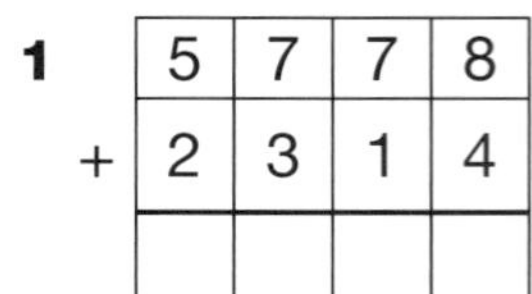

2
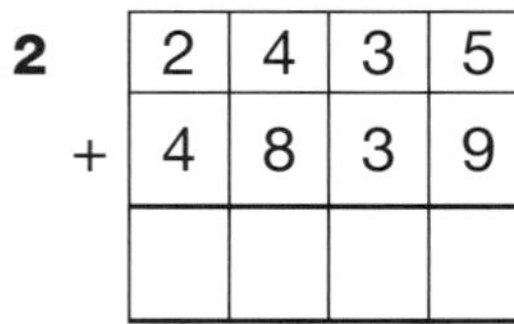

3
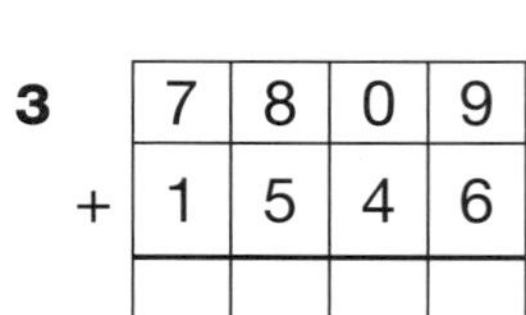

4
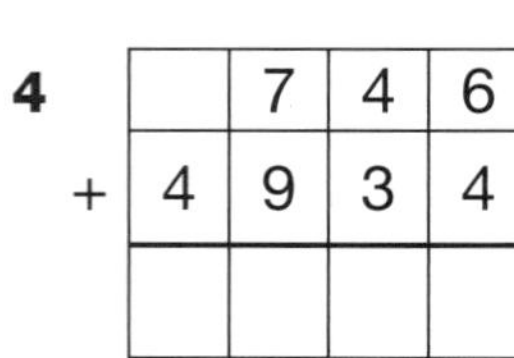

5
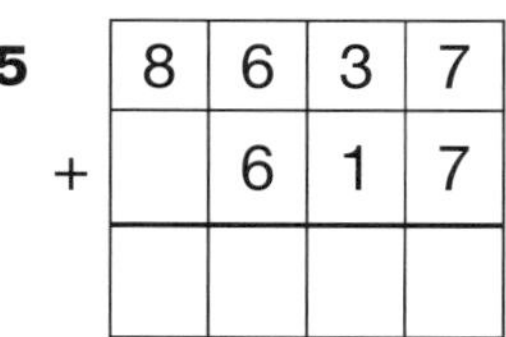

6
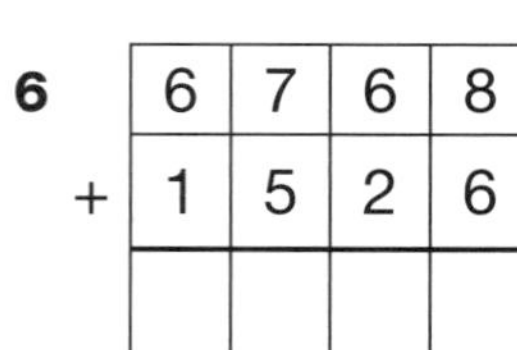

Working Mathematically

7 Round each number to the nearest 1000 to estimate how many people attended the music festival in total if 4988 went on Saturday and 5012 went on Sunday.

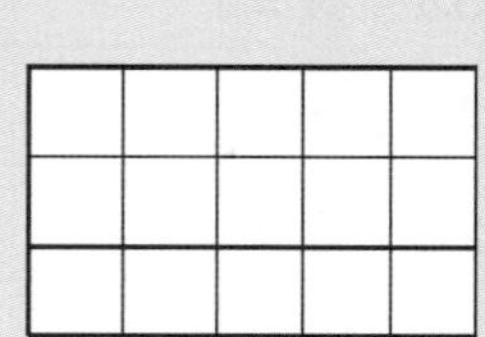

Number Addition and subtraction

1 Milton's dishwasher was priced at \$3465, but he received a discount of \$500. How much did he pay?

			=	

2 Navaya has already driven 1236 km, but she still has another 965 km to go. What will be her total distance travelled?

			=	

3 What was the total number of tickets sold, if 4165 tickets were sold on Saturday and 3535 were sold on Sunday?

4 On 21 April, Rumi had \$3588 in her bank account and on 30 April she deposited \$1050. How much is in the account now?

Number and Algebra

SET 3 Tenths and hundredths

Complete the table.

	Tenths	Hundredths	Decimal
1	$\frac{1}{10}$		0.10
2	$\frac{3}{10}$		0.30
3		$\frac{90}{100}$	0.90
4	$\frac{5}{10}$		0.50
5	$\frac{7}{10}$		0.70
6		$\frac{20}{100}$	0.20
7	$\frac{6}{10}$	$\frac{60}{100}$	

Write true or false.

8 $\frac{3}{10} = 0.3$ ______________

9 $\frac{7}{10} = \frac{70}{100}$ ______________

10 $0.5 = \frac{5}{10}$ ______________

11 $\frac{2}{10} > \frac{5}{10}$ ______________

12 $0.3 < 0.7$ ______________

13 $\frac{30}{100} = \frac{3}{10}$ ______________

14 $0.6 < 0.2$ ______________

15 $\frac{5}{10} > \frac{7}{10}$ ______________

16 $\frac{40}{100} < 0.5$ ______________

SET 4 Extension

1 $92 – $13

2 4 L = ☐ mL

3 List the factors of 36.

4 Quarter of a litre

5 Share 50c among 10 people.

6 How many hundredths in a whole?

7 6:15 am + 20 mins

8 $6^2 - 5$

9 $(6 \times 9) - 4$

10 List the factors of 24.

11 How much is $3\frac{1}{2}$ kg of meat at $1.20 kg?

12 What is the product of 8 and 6?

Working Mathematically

13 Use rounding strategies to estimate who trains for the longer time.

Sally trains for 57 minutes 5 days a week. Rod trains for 39 minutes 7 days a week.

Use a calculator to check your estimates.

Statistics and Probability Recording data

	A	B	C	D
1	Date	Item	Cost	Balance
2	1 June	opening		$400
3	7 June	food	$160	(D2 – C3) $240
4	10 June	petrol	$40	(D3 – C4) $200
5	11 June	bills	$120	(D4 – C5) $80
6	20 June	drinks	$30	(D5 – C6) $50
7	29 June	movies	$20	(D6 – C7) $30

1 How much was spent on food?

2 How much was spent on bills?

3 How much was spent on movies?

4 What was the date when the balance equalled $200?

5 What was the date when the balance equalled $50?

6 Would there be enough money on 30 June to buy a radio costing $55?

UNIT 31

Number and Algebra

SET 1 Basic

1 16 – 5

2 6 × 8

3 6 + 8

4 7 × 8

5 8 × 4

6 10 + 10 + 8 – 18

7 What is the sum of 18 and 12?

8 What is the product of 9 and 9?

9 How many threes in 24?

10 Half of $26

11 $13.50 = ☐ c

12 2706 = ☐ thousands + ☐ hundreds + ☐ tens + ☐ ones

13 What is the difference between 36 and 14?

14 (3 × 7) + 9

15

I have 21 points. How many more do I need to equal the record, which is 57?

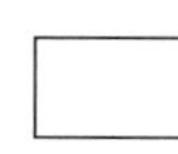

SET 2 Multiplication

1 What is 4 lots of 8?

2 Does 8 × 4 equal 4 × 8?

3 Does 7 × 10 equal 10 × 7?

4 What is 6 lots of 7?

5 Add 7 ten times.

6 $\begin{array}{r} 2\ 3 \\ \times \quad 4 \\ \hline \end{array}$

7 $\begin{array}{r} 2\ 6 \\ \times \quad 6 \\ \hline \end{array}$

8 $\begin{array}{r} 2\ 5 \\ \times \quad 5 \\ \hline \end{array}$

9 $\begin{array}{r} 3\ 5 \\ \times \quad 3 \\ \hline \end{array}$

10 $\begin{array}{r} 6\ 5 \\ \times \quad 5 \\ \hline \end{array}$

11 $\begin{array}{r} 5\ 7 \\ \times \quad 2 \\ \hline \end{array}$

12 Marco earns $20 per hour on the weekend and $15 per hour during the week. How much did he earn if he worked 6 hours on the weekend and 8 hours during the week?

Space Nets

Match the following objects to their nets.

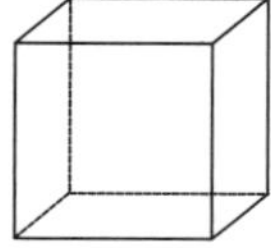

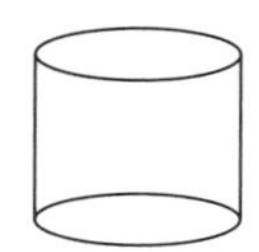

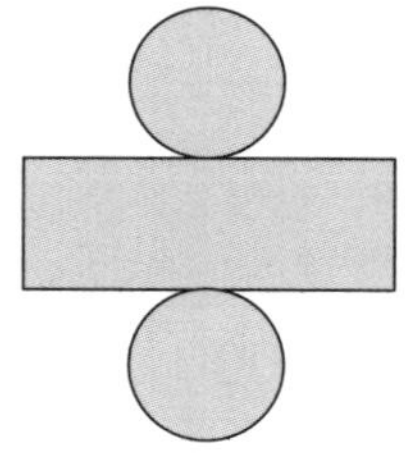

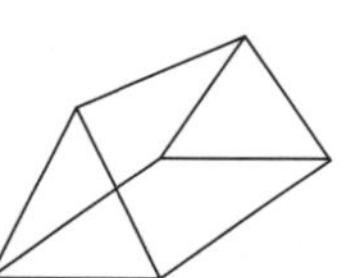

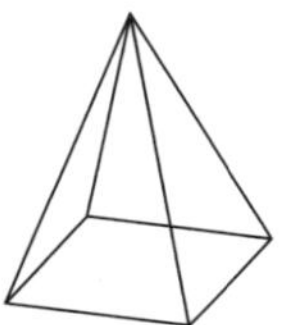

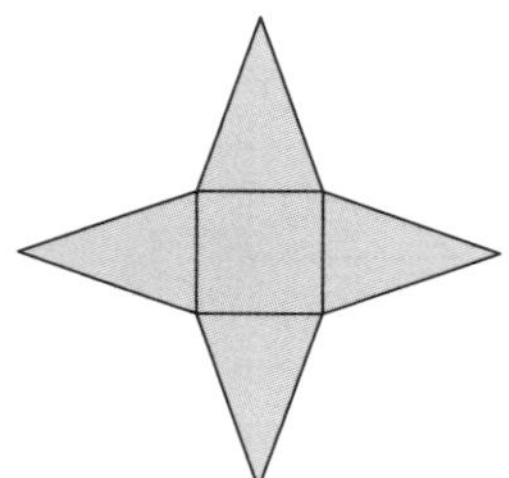

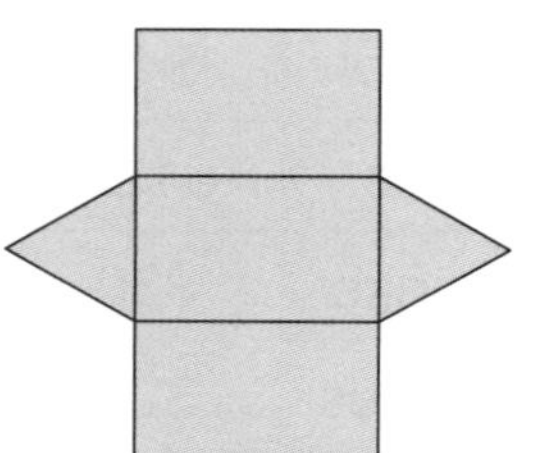

Number and Algebra

SET 3 Decimals

Shade the largest decimal.

1	0.5	0.7	0.2
2	0.6	0.9	0.3
3	0.34	0.58	0.41
4	2.3	2.75	2.12
5	5.95	4.62	5.34
6	7.62	6.84	8.11

1.85 m 1.27 m 0.7 m 4.8 m

Round each height to the nearest metre.

7 Refrigerator ______ m

8 Chair ______ m

9 Pot plant ______ m

10 Flag pole ______ m

11 Is the chair taller than the refrigerator? ____

SET 4 Extension

1 How many grams in $1\frac{1}{4}$ kg?

2 Share 42 among 3 people.

3 Five books at $41 each

4 What is the value of 1 in 3015?

5 200 mL = 2 L. True or false?

6 Write the smallest number you can using 1, 8, 7, 3.

7 How many tenths in 2 wholes?

8 37 hundredths = 0. ☐

9 Write 1 whole and 3 tenths as a decimal.

10 How many centimetres in 3.25 m?

11 Is 128 a multiple of 7?

12 If 6 cakes cost $36, how much for 11?

13 How much is $2\frac{1}{2}$ kg of chicken at $6 a kilogram?

Working Mathematically

14 Jessica bought two items. If the phone cost $15 less than the stereo, how much change would she receive from $200?

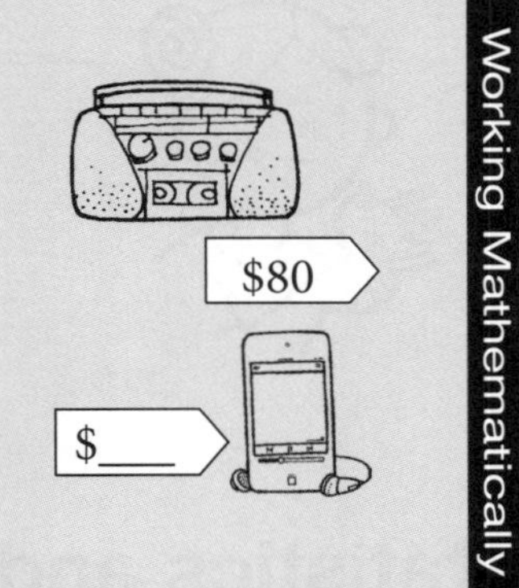

Measurement Decimal notation

Place the athletes' names on the victory board according to how far they jumped in the long jump competition.

Jess	2.41 metres
Caleb	3.06 metres
Eboni	2.52 metres
Jack	2.96 metres
Henri	3.27 metres

4th	2nd	1st	3rd	5th

UNIT 32

Number and Algebra

SET 1 Basic

1 6 × 5

2 What is the product of 6 and 4?

3 3 mL + 10 mL

4 Triple 10c.

5 $1.70 – 20c

6 228 cm = [ ] m [ ] cm

7 10 kg – 3 kg

8 Square 8.

9 21 m ÷ 3

10 What is the product of 9 and 9?

11 $11 – $10.50

12 42 m ÷ 6

13 20 mL + 25 mL

14 21 × 4

15

What number is it? It's less than 20. It's more than 3 × 4, and one of its digits is 6.

SET 2 Number patterns

Follow the rules to complete the number patterns.

1

Multiply by 5							
2	4	6	8	10	12	14	16

2

Add 17							
1	2	3	4	5	6	7	8

3

Multiply by 4 then add 1							
1	2	3	4	5	6	7	8

4

Multiply by 6 then take away 2							
3	4	5	6	7	8	9	10

Use the rule and complete the pattern.

5

Multiply by 4 then subtract 3							
1	2	3	4	5	6	7	8
1	5	9					

6

Multiply by 7 then add 4							
1	2	3	4	5	6	7	8
11	18	25					

Statistics and Probability Investigating graphs

Ms Sunder's class surveyed and graphed the points scored by 6 teams.

1 Which teams are leading the competition?

____

2 Which team is coming last?

____

3 Which team has 200 points?

____

4 How many more points does GP need to join the leaders?

____

Basketball points table

Number and Algebra

SET 3 Fraction/decimal patterns

Tick the correct patterns.

1	0.1	0.2	0.3	0.4	0.5	0.6	0.7	
2	$\frac{1}{2}$	1	$1\frac{1}{2}$	2	$2\frac{1}{2}$	$3\frac{1}{2}$	$4\frac{1}{2}$	
3	$\frac{2}{10}$	$\frac{4}{10}$	$\frac{6}{10}$	$\frac{8}{10}$	1	$1\frac{2}{10}$	$1\frac{4}{10}$	
4	0.5	0.6	0.7	0.8	1	1.2	1.4	
5	$\frac{1}{4}$	$\frac{2}{4}$	$\frac{3}{4}$	1	$1\frac{1}{4}$	$1\frac{2}{4}$	$1\frac{3}{4}$	
6	1.15	1.20	1.25	1.30	1.35	1.40	1.45	

Complete these patterns.

7

Add $\frac{1}{4}$					
0	$\frac{1}{4}$	$\frac{2}{4}$	$\frac{3}{4}$	1	$1\frac{1}{4}$
$\frac{1}{4}$	$\frac{2}{4}$				

8

Add 0.2					
1	1.2	1.4	1.6	1.8	2
1.2	1.4				

SET 4 Extension

1 (3 × 7) + 0

2 (3 × 6) + 12 = 30. True or false?

3 Round 4506 to the nearest 1000.

4 Centimetres in 4.27 m?

5 What is the value of 8 in 11 874?

6 Write 7 tenths as a decimal.

7 Write 17 hundredths as a decimal.

8 If 9 apples cost $8.10, how much would 5 cost?

9 Is 48 a multiple of 9?

10 John's car averaged 100 km/h over $4\frac{1}{2}$ hours. How far did it travel?

11 Write all the factors of 35.

12 Write 7 hundredths as a decimal.

13 Complete the path.

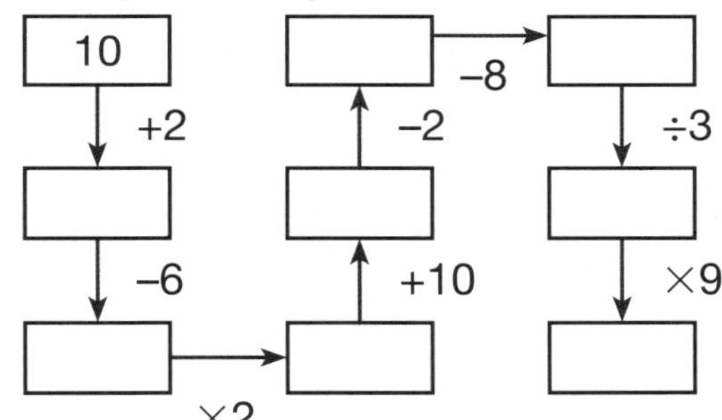

Measurement The square metre

Tarun was retiling his kitchen/family room. He drew square metres on the floor to calculate how many tiles to order.

1 How many square metres of tiles does he need to cover his kitchen benches?

m²

2 What is the area of the whole kitchen/family room?

m²

Kitchen/family room

Benches

Kitchen

Fridge

Number and Algebra

SET 1 Basic

1 3 × 6

2 6 + 6 + 8

3 3 × 8

4 17 – 6

5 45 – 6

6 18 + 4

7 3 × 9

8 3 × 7

9 What is the product of 7 and 5?

10 How many days in August?

11 What is the sum of 39 and 6?

12 $31 – $7

13 What is the difference between 37 and 20?

14 8756 = ☐ thousands + ☐ hundreds + ☐ tens + ☐ ones

15

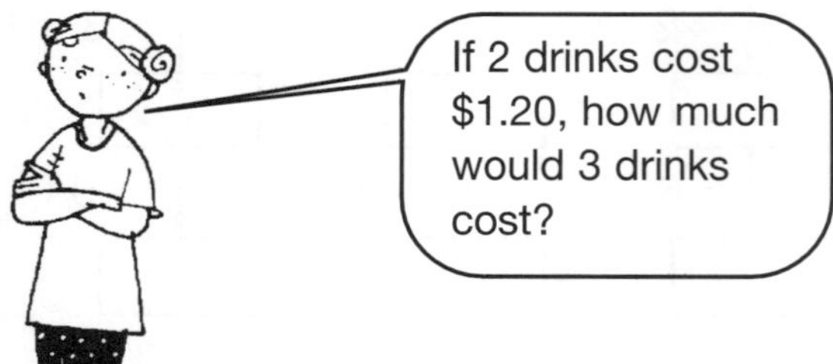

SET 2 Place value

Multiply each number in the table by 10, 100 and 1000.

	Number	× 10	× 100	× 1000
1	5			
2	8			
3	19			
4	27			
5	45			
6	61			
7	77			
8	80			
9	1			
10	99			

11 How many tens are in 487? ________

12 How many hundreds are in 555? ________

13 How many hundreds are in 2155? ________

Use the less than < or greater than > sign to compare these numbers.

14 5999 ☐ 6000

15 9595 ☐ 5959

Statistics and Probability Representing data

Create a column graph to represent the data in the table.

Favourite sports	
Netball	𝍸 𝍸 𝍸
Football	𝍸 𝍸 𝍸 𝍸 |
Gymnastics	𝍸 |||
Swimming	𝍸 𝍸 |||
Athletics	𝍸 ||||

Favourite sports

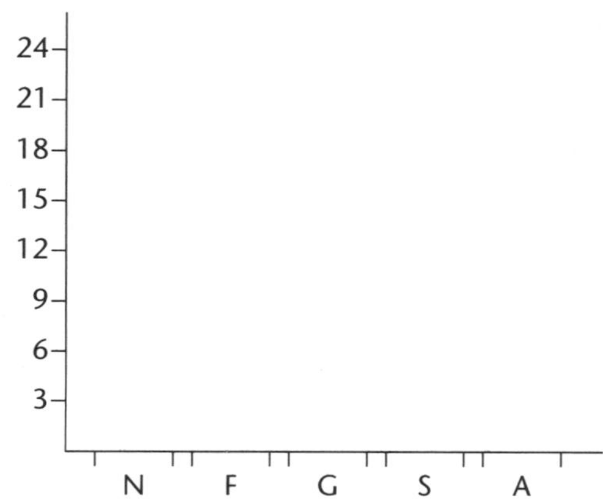

Number and Algebra

SET 3 Associative property

Rewrite these number sentences to demonstrate that it doesn't matter in which order numbers are added or multiplied.

1 $7 + 9 + 3 = 19$

2 $6 + 8 + 4 = 18$

3 $18 + 5 + 2 = 25$

4 $5 + 6 + 9 = 20$

5 $3 \times 2 \times 5 = 30$

6 $6 \times 5 \times 2 = 60$

7 $3 \times 5 \times 4 = 60$

Solve the number sentences.

8 $3 \times 7 + 2 =$

9 $8 \times 7 + 4 =$

10 $5 \times 2 \times 3 + 20 =$

11 $9 \times 8 - 3 =$

12 $3 \times 4 \times 6 =$

13 $7 + 5 + 2 =$

SET 4 Extension

1 $\frac{1}{4}$ of 56

2 $(8 \times 7) - 6$

3 Share 98 stickers among 6.

4 What numeral is in the thousands place in 7821?

5 3000 g = ☐ kg

6 Write the numeral for seven thousand and ten.

7 List the factors of 32.

8 $5679 - 137$

9 How many edges has a cube?

10 Is 36 a multiple of 9?

11 How much is 7 kg of bacon at $3.50 per kilogram?

12 Round 6328 to the nearest 100.

13 Estimate an answer to 98×5.

Working Mathematically

14 Mahira had $480 in the bank and Paco had half what Mahira had plus $15. What was the total amount they had altogether?

Measurement Litres and millilitres

Use decimal notation to record the amount of water in each container.

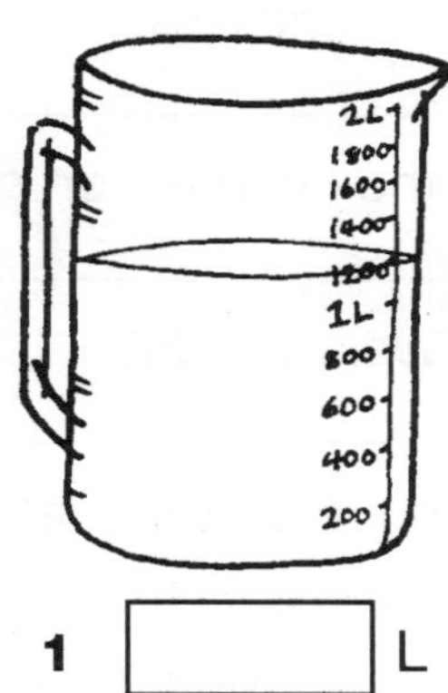

1 ☐ L

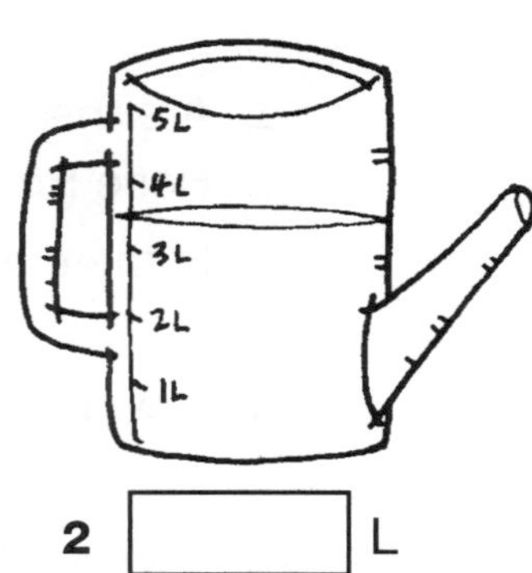

2 ☐ L

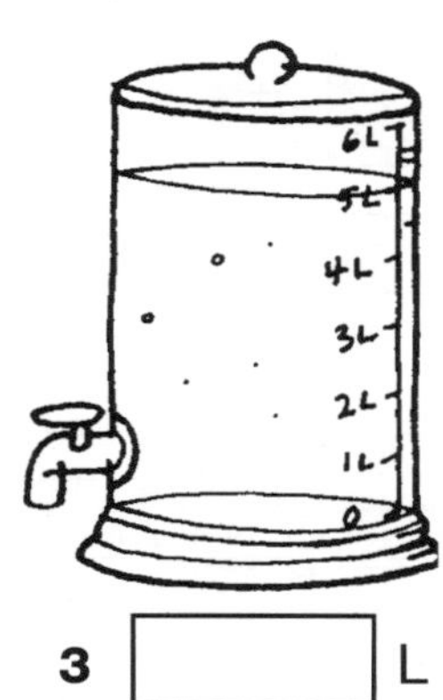

3 ☐ L

UNIT 34

Number and Algebra

SET 1 Basic

1 3 × 5

2 48 – 5

3 17 + 8

4 32 ÷ 4

5 5 × 8

6 20 ÷ 5

7 What is the sum of 13 and 11?

8 What is the difference between 43 and 41?

9 Triple 8.

10 What is half of $42?

11 $16.03 = ☐ c

12 6^2

13 

The party started at a quarter to four. We were ten minutes late. When did we get there?

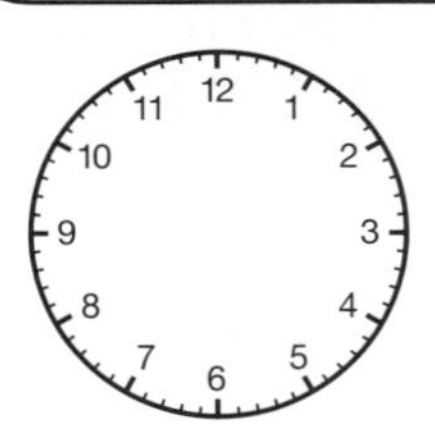

SET 2 Multiplication facts

Supply the missing numbers so that both sides of the balance beam are equal.

1 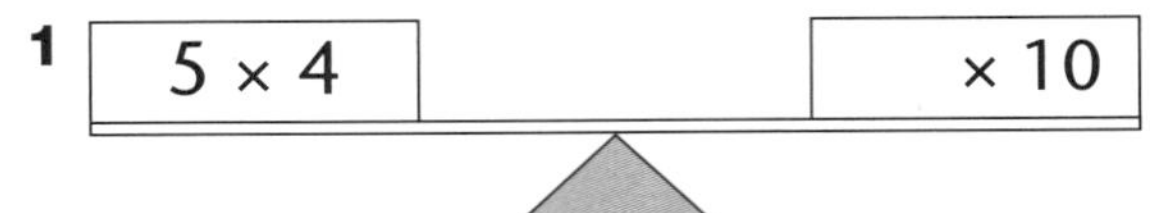

2 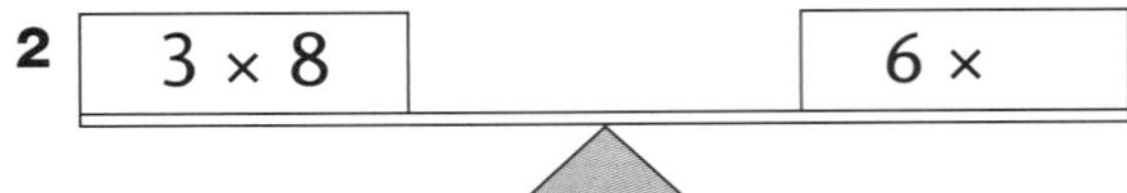

3 6 × 5 | 15 ×

4 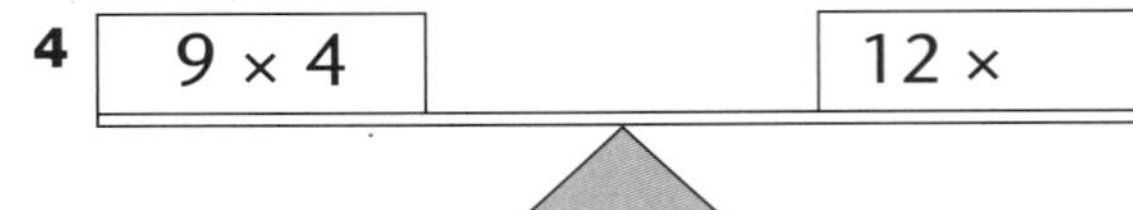

5 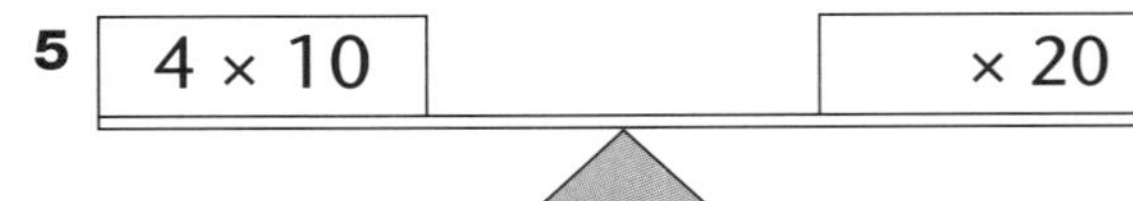

6 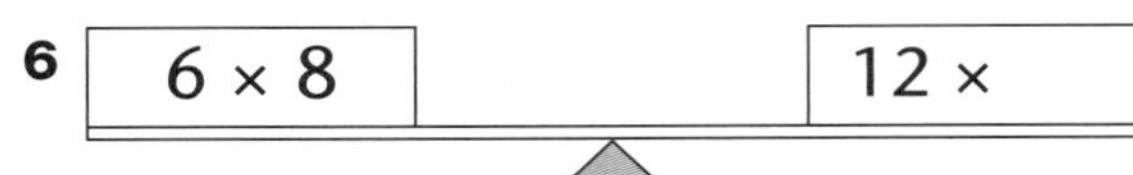

7 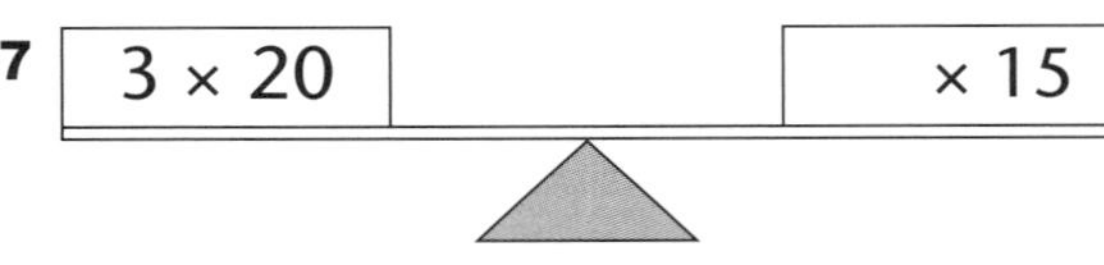

Space Using a legend

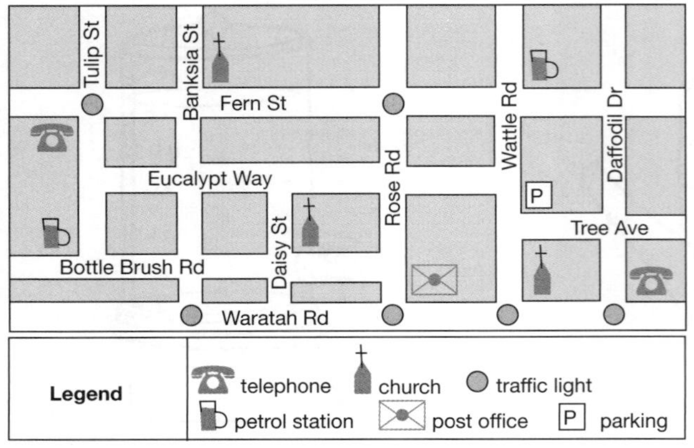

1 How many churches are on the map?

2 Are there traffic lights at the intersection of Rose Rd and Fern St?

3 Is there a post office on Waratah Rd?

4 How many petrol stations are on the map?

5 Name a street that has a telephone booth.

Number and Algebra

SET 3 Multiplication

1 What number has the factors 7 and 8?

2 What number has the factors 9 and 5?

3 Triple 9.

4 What number is 6 times larger than 8?

5 5 squared

6 What number is 4 times larger than 10?

7 How many sevens are needed to make 49?

8

	HUND	TENS	ONES
		3	5
×			4

9

	HUND	TENS	ONES
		4	5
×			5

10

	HUND	TENS	ONES
		8	4
×			6

11

	HUND	TENS	ONES
		8	4
×			7

Working Mathematically

12 Jim's sister spilt ink on his homework. Write in the missing numbers to help Jim.

	HUND	TENS	ONES
		^{2}5	
×			6
		2	4

SET 4 Extension

1 Write the number that is 5 before 303.

2 (8 × 7) + 6

3 111, 131, 151, ☐

4 Subtract 70 from 3386.

5 How many corners has a triangular prism?

6 Write 8 tenths as a decimal.

7 How much are 9 apples at 45c each?

8 Write 4107 in words.

____

____

9 How many hours in 6 days?

10 Is 36 a multiple of 7?

11 How much is $4\frac{1}{2}$ kg of potatoes at 80c per kilogram?

12 Write the set of factors for 20.

13 Multiply $136 by 7.

14 4137, 4146, 4155, ☐

15 How many halves in $5\frac{1}{2}$?

16 The taxi made 9 trips. There were 45 passengers. What was the average number of passengers for each trip?

Measurement Displacement

How much water was displaced by each rock if there was 500 mL of water in each jug to begin?

1

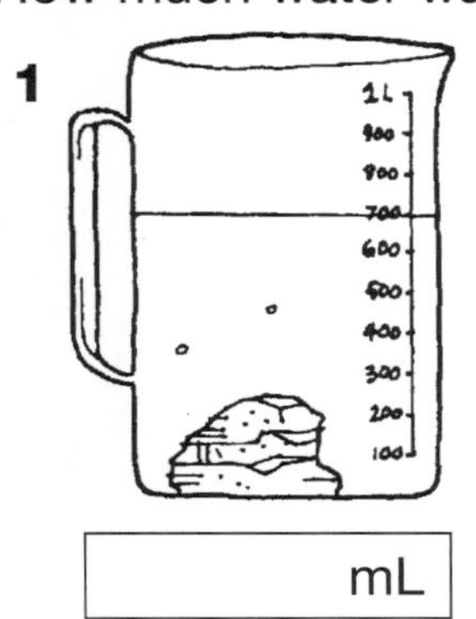

☐ mL

2

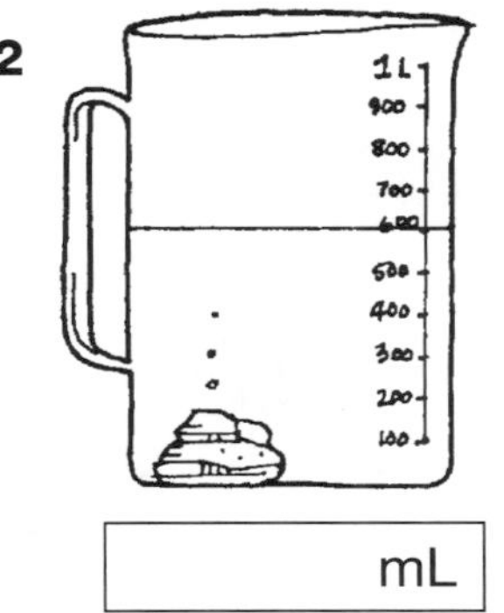

☐ mL

3

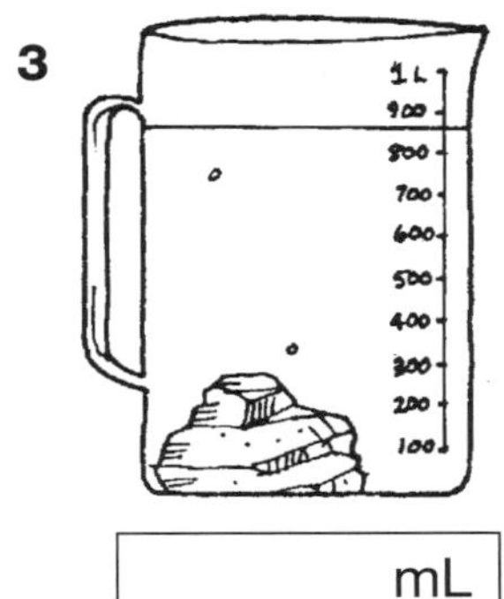

☐ mL

4

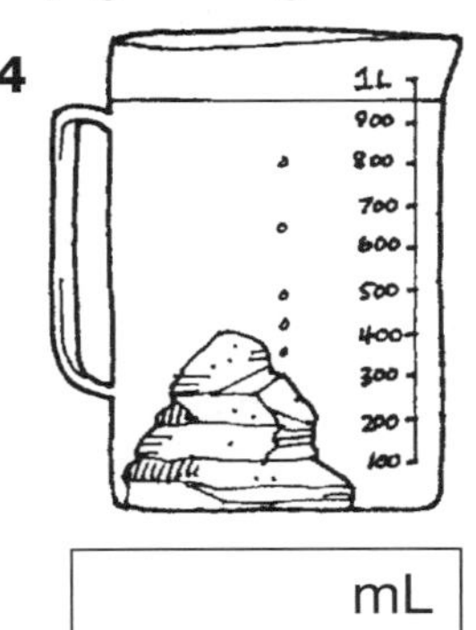

☐ mL

UNIT 35

Number and Algebra

SET 1 Basic

1 $6 × 7

2 Share $32 among 8 people.

3 What is the difference between 20 and 9?

4 $16 × 2

5 What is the sum of 14, 18 and 20?

6 What is the difference between 90 and 39?

7 $6 + $7 + $7

8 $14.65 = ☐ c

9 What is the product of 7 and 8?

10 Triple 9.

11 13 × 3

12 15 kg – 9 kg

13 What is the difference between 48 and 30?

14 (2 + 1) × 2

15 90, 80, ☐, ☐, 50, 40, ☐

16

Jane had 40 marbles but lost 7 and gave away 13. How many did she have left? ☐

SET 2 Subtraction

1

	THOU	HUND	TENS	ONES
	7	7	4	7
–	3	4	3	3

2

	THOU	HUND	TENS	ONES
	6	8	3	2
–	2	6	1	3

3

	THOU	HUND	TENS	ONES
	7	7	4	0
–	3	4	1	7

4

	THOU	HUND	TENS	ONES
	3	6	9	0
–	1	2	3	4

5

	THOU	HUND	TENS	ONES
	8	5	2	6
–	3	1	6	2

6

	THOU	HUND	TENS	ONES
	7	5	3	7
–	2	3	4	8

7 What is the difference between 2500 and 3000?

8 What is the difference between 1250 and 2000?

9 Deduct $700 from $5000.

10 Deduct $25 from $3250.

Space Isometric drawings

Draw the object on the isometric dot paper. The tops have been drawn for you.

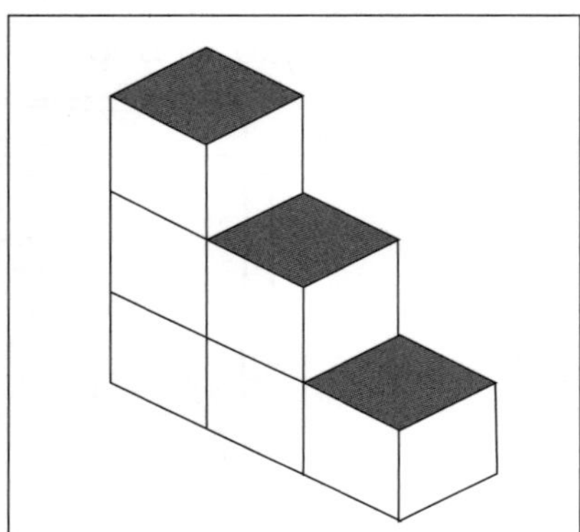

Number and Algebra

SET 3 Odd and even numbers

Answer the questions.

1 If you add two odd numbers can you get a total which is an odd number? ________

2 If you add two even numbers can you get a total which is an odd number? ________

3 If you add an even number and an odd number can you get a total which is an odd number? ________

4 If you take away an odd number from another odd number, can you get an answer which is an odd number?

____

5 If you multiply two even numbers can you get a product which is an odd number? ________

6 If you multiply an even number by an odd number can you get a product which is an odd number? ________

7 If you multiply two odd numbers can you get a product which is an odd number? ________

SET 4 Extension

1 $(4 \times 8) + 5$

2 How much are 10 pens at $4.50 each?

3 How many centimetres in $3\frac{3}{4}$ m?

4 Is 21 a multiple of 3?

5 How much would each person pay if 4 players hired a tennis court at $36 per court?

6 Multiply the sum of 5 and 4 by 7.

7 How much is $3\frac{1}{2}$ kg of meat at $7 per kilo?

8 Round to the nearest 100 to estimate an answer to 698 + 717.

9 Lisa drank 5 mL of medicine at 8 am, then 5 mL every 4 hours. How much did she drink between 8 am and 9 pm?

Working Mathematically

10 A rectangle has a perimeter of 34 m and an area of 72 m^2. What are the length and width of its sides?

Length ________ Width ________

Space Position coordinates

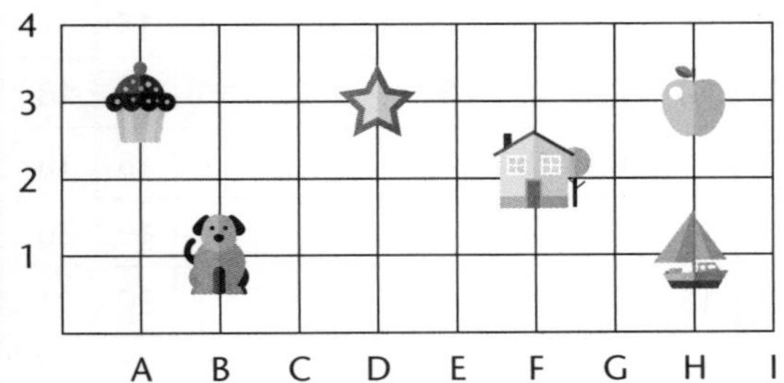

Record the coordinates of each item.

	Item	Coordinate
1	(star)	
2	(cupcake)	
3	(boat)	

	Item	Coordinate
4	(apple)	
5	(house)	
6	(dog)	

Maths helpers

Length

10 millimetres (mm) = 1 centimetre (cm)

100 centimetres (cm) = 1 metre (m)

1000 metres (m) = 1 kilometre (km)

Mass

1000 grams (g) = 1 kilogram (kg)

1000 kilograms (kg) = 1 tonne (t)

Capacity

1000 millilitres (mL) = 1 litre (L)

Time

60 seconds = 1 minute

60 minutes = 1 hour

24 hours = 1 day

7 days = 1 week

14 days = 1 fortnight

12 months = 1 year

52 weeks = 1 year

365 days = 1 year

366 days = 1 leap year

10 years = 1 decade

100 years = 1 century

Months of the year

Thirty days has September, April, June and November. All the rest have thirty-one, except February alone, which has twenty-eight days clear and twenty-nine days each leap year.

Seasons

Summer: December, January, February

Autumn: March, April, May

Winter: June, July, August

Spring: September, October, November

Roman numerals

1 = I	30 = XXX
2 = II	40 = XL
3 = III	50 = L
4 = IV	60 = LX
5 = V	70 = LXX
6 = VI	80 = LXXX
7 = VII	90 = XC
8 = VIII	100 = C
9 = IX	500 = D
10 = X	1000 = M
20 = XX	

Multiplication facts

×	0	1	2	3	4	5	6	7	8	9	10
0	0	0	0	0	0	0	0	0	0	0	0
1	0	1	2	3	4	5	6	7	8	9	10
2	0	2	4	6	8	10	12	14	16	18	20
3	0	3	6	9	12	15	18	21	24	27	30
4	0	4	8	12	16	20	24	28	32	36	40
5	0	5	10	15	20	25	30	35	40	45	50
6	0	6	12	18	24	30	36	42	48	54	60
7	0	7	14	21	28	35	42	49	56	63	70
8	0	8	16	24	32	40	48	56	64	72	80
9	0	9	18	27	36	45	54	63	72	81	90
10	0	10	20	30	40	50	60	70	80	90	100

Addition facts

+	2	3	4	5	6	7	8	9	10	11	12
2	4	5	6	7	8	9	10	11	12	13	14
3	5	6	7	8	9	10	11	12	13	14	15
4	6	7	8	9	10	11	12	13	14	15	16
5	7	8	9	10	11	12	13	14	15	16	17
6	8	9	10	11	12	13	14	15	16	17	18
7	9	10	11	12	13	14	15	16	17	18	19
8	10	11	12	13	14	15	16	17	18	19	20
9	11	12	13	14	15	16	17	18	19	20	21
10	12	13	14	15	16	17	18	19	20	21	22
11	13	14	15	16	17	18	19	20	21	22	23
12	14	15	16	17	18	19	20	21	22	23	24

Answers

UNIT 1 Number and Algebra

SET 1

1 12
2 17
3 4
4 15
5 30
6 28
7 20
8 March
9 14
10 3
11 6
12 7c
13 15
14 $7

SET 2

1 50, 70, 30, 60
2 80, 60, 70, 90
3 80, 60, 100, 120
4 100, 120, 80, 140
5 9
6 90
7 900
8 130
9 1300
10 120 books
11 140 books
12 230 books

SET 3

1 $15
2 $18
3 $15
4 $21
5 $24
6 $27
7 $30
8 10, 35, 20, 30, 45, 40
9 20, 70, 40, 60, 90, 80
10 Hands on. (Second column × 10 is double the first column × 5.)

SET 4

1 317
2 265
3 450
4 159
5 600
6 $18
7 208
8 359
9 190
10 24
11 $54
12 41
13 37
14 8
15 Four thousand, two hundred and seven
16 1, 2, 3, 4, 6, 8, 12, 16, 24, 48
17 0.3
18 3900

Space

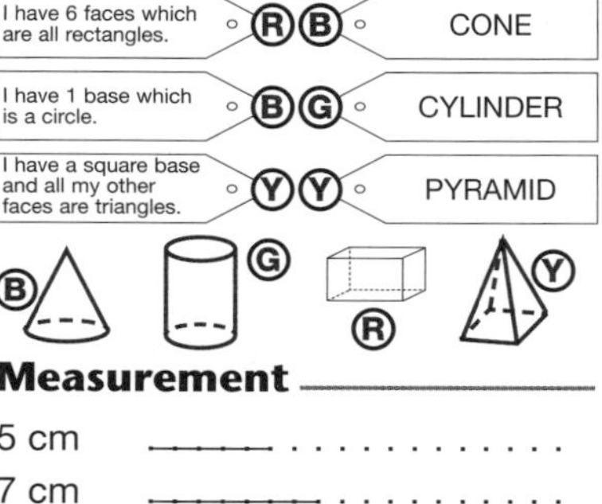

Measurement

5 cm
7 cm
9 cm
13 cm
$14\frac{1}{2}$ cm
$15\frac{1}{2}$ cm

UNIT 2 Number and Algebra

SET 1

1 7
2 3
3 8
4 4
5 20
6 6
7 24
8 4
9 February
10 18
11 22
12 22
13 120
14 4 thousands + 3 hundreds + 7 tens + 2 ones
15 11

SET 2

1 58
2 67
3 39
4 57
5 48
6 47
7 406
8 309
9 444
10 $34
11 49

SET 3

1 seven hundreds
2 four tens
3 three thousands
4 six ones
5 five hundreds
6 856
7 2457
8 4576
9 2775
10 3474
11 397
12 1397
13 3540
14 3464
15 1296
16 753
17 357

SET 4

1 19
2 34
3 Sunday
4 40, 35, 30
5 9
6 563
7 7:45 pm
8 $1.15
9 $7.20
10 75c
11 9864
12 Possible answers include:
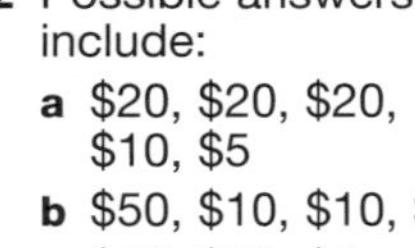
a $20, $20, $20, $10, $5
b $50, $10, $10, $5
c $50, $20, $5

Space

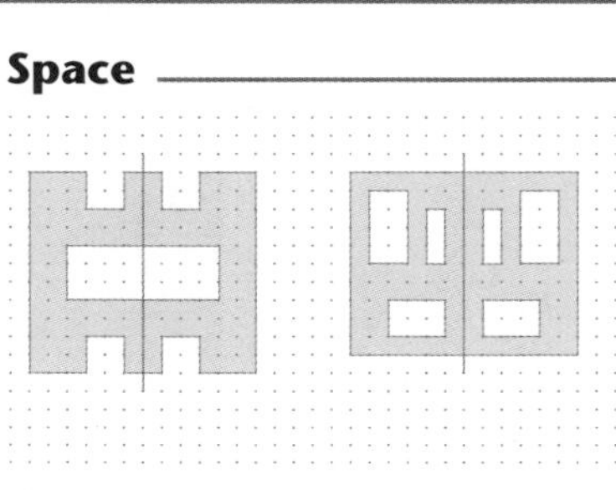

Measurement

1 10 cm^2
2 14 cm^2
3 18 cm^2

UNIT 3 Number and Algebra

SET 1

1 19
2 2
3 8
4 10
5 14
6 24
7 7
8 18
9 5
10 63
11 9
12 15
13 35
14 18
15 $13

SET 2

1 81
2 45
3 98
4 92
5 62
6 81
7 851
8 777
9 81
10 AUSTRALIA

SET 3

1 20, 25, 30, 35, 40
2 9, 12, 15, 18, 21
3 50, 60, 70, 80, 90
4 32, 35, 38, 41, 44
5 222, 224, 226, 228, 230
6 310, 305, 300, 295, 290
7 350, 300, 250, 200, 150
8 700, 750, 800, 850, 900
9 845
10 4231
11 3707

SET 4

1 127
2 F
3 6
4 9 each
5 2000
6 200
7 62
8 3206
9 2497
10 1300
11 1378
12 $12
13 12
14 28
15 One thousand, four hundred
16 $2
17 About 250

Space

Woman, bottom left corner.

Statistics and Probability

1 7
2 8
3 Toy cars
4 Books
5 28
6 Games and toy cars

Answers

UNIT 4 Number and Algebra

SET 1

1 9
2 10
3 3
4 2
5 15
6 24
7 20
8 30
9 January
10 9
11 5
12 7
13 60
14 46
15 5

SET 2

1 22
2 26
3 18
4 45
5 15
6 21
7 15
8 13
9 33
10 47
11 31
12 42
13 35

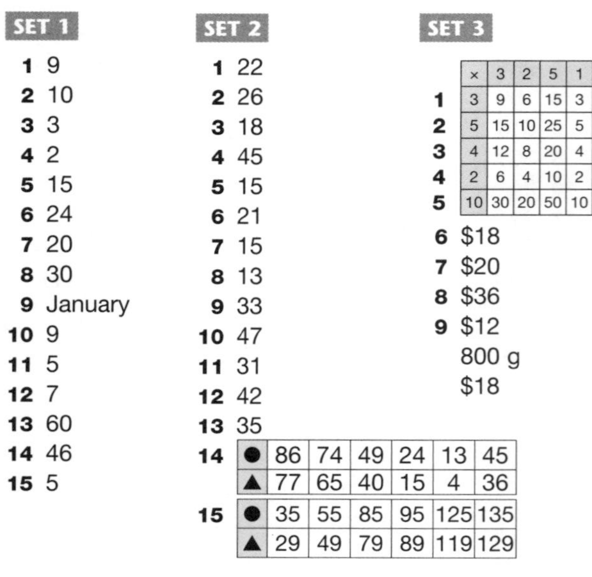

14

●	86	74	49	24	13	45
▲	77	65	40	15	4	36

15

●	35	55	85	95	125	135
▲	29	49	79	89	119	129

SET 3

	×	3	2	5	1	0	7	9
1	3	9	6	15	3	0	21	27
2	5	15	10	25	5	0	35	45
3	4	12	8	20	4	0	28	36
4	2	6	4	10	2	0	14	18
5	10	30	20	50	10	0	70	90

6 $18
7 $20
8 $36
9 $12
800 g
$18

SET 4

1 H
2 500
3 $8 each
4 $9
5 $4.50
6 80
7 70
8 280
9 No (5)
10 $21
11 Possible answers include:
50 + 30 + 20 = 100
50 + 10 + 40 = 100
40 + 40 + 20 = 100

Space

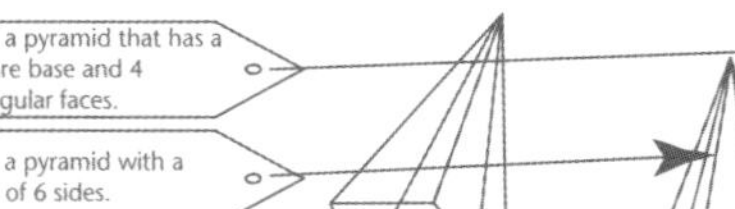

Measurement

	Estimate	cm
1	Hands on	10 cm
2	Hands on	11 cm
3	Hands on	12 cm
4	Hands on	9 cm
5	Hands on	7 cm
6	Hands on	4 cm

UNIT 5 Number and Algebra

SET 1

1 11
2 10
3 14
4 20
5 18
6 8
7 20
8 18
9 June
10 20
11 4
12 6
13 14
14 180
15 Five thousand, two hundred and seven

SET 2

1 24
2 240
3 18
4 180
5 32
6 320
7 28
8 280
9 14
10 140
11 25
12 250

SET 3

1 90
2 60
3 180
4 490
5 60 + 30 = 90
6 70 + 30 = 100
7 50 + 30 = 80
8 70 + 40 = 110
9 100 + 50 = 150
10 80 – 50 = 30
11 70 – 40 = 30
12 90 – 60 = 30
13 60 – 20 = 40
14 80 – 80 = 0

SET 4

1 32
2 5 each, remainder 1
3 4825
4 $1.80
5 600
6 17
7 $6.10
8 $135
9 6
10 12
11 $28
12 400
13 $7 each
14 396, 393, 390
15 5

Space

1 Hands on
2

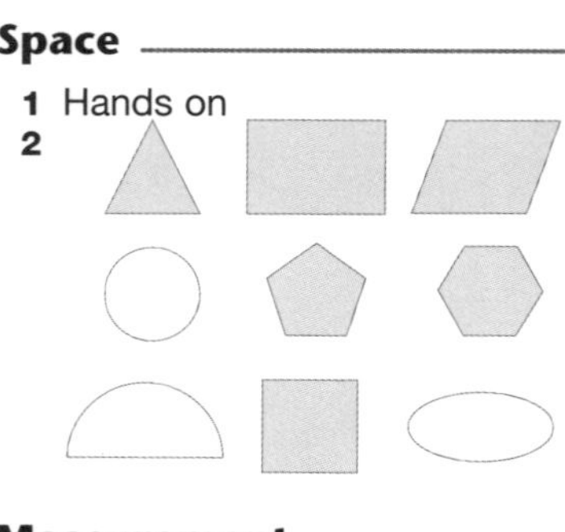

Measurement

1 60
2 60
3 24
4 7
5 14
6 365, 366 (leap year)
7 hours/minutes
8 hours/minutes
9 minutes
10 seconds

UNIT 6 Number and Algebra

SET 1

1 29
2 80
3 15
4 21
5 24
6 12
7 15
8 16
9 15
10 18
11 13
12 9
13 24
14 8
15 7637
16 $18

SET 2

1
3 thousands 7 hundreds 4 tens 3 ones
3 7 hundreds 4 tens 3 ones
3 7 4 tens 3 ones
3 7 4 3 ones

2 3
3 13
4 124
5 2
6 32
7 25
8 3
9 3668
10 1367
11 8060
12 3588
13 2401
14 6070
15 1556

SET 3

1 18
2 30
3 42
4 54
5 12
6 24
7 36
8 48
9 60
10 15
11 35
12 27
13 30
14 9
15 8
16 3
17 50
18 80
19 a 30 mL
b 180 mL

SET 4

1 95
2 85
3 100
4 20
5 9
6 30c
7 950
8 22
9 12
10 5
11 4
12 2640
13 48
14 4085
15 1324
16

Space

1 A2
2 D3
3 G1
4 B5
5 F4
6 G6
7–8 Hands on

Statistics and Probability

1 4
2 8
3 11
4 6 years
5 6 years

UNIT 7 Number and Algebra

SET 1

1 14
2 16
3 13
4 11
5 40
6 45
7 21
8 12
9 33
10 37
11 21
12 April
13 15
14 27
15 $3237

SET 2

1 6
2 4
3 8 each
4 8
5 3
6 6
7 10 each
8 9 each
9 7 each
10 3
11 9
12 7
13 6
14 8
15 7

SET 3

1 60
2 90
3 184
4 36
5 36
6 28
7 29
8 20
9 200
10 200
11 412 cm
12 12, 18, 6, 24, 30

SET 4

1 120
2 331
3 500
4 2800
5 30
6 164
7 $7 each
8 14
9 315
10 16
11 40
12 53
13 3 at $1.10 each and $1 + $1.20 + $1.10

Space

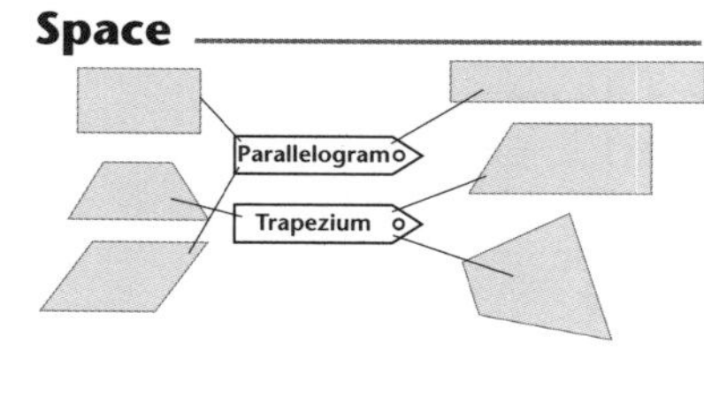

Measurement

1 milk carton
2 cordial bottle
3 juice bottle
4 bucket
5 oil drum

UNIT 8 Number and Algebra

SET 1

1 13
2 10
3 16
4 16 km
5 18
6 28
7 20
8 0
9 May
10 19
11 $2 each
12 24
13 10c
14 4:18
15 643
16 895

SET 2

1 8, 16, 20, 24, 28, 32
2 14, 21, 28, 42, 56, 63
3 12, 24, 30, 36, 54, 48
4 $42
5 $21
6 $72
7 $56
8 41
9 90
10 70
11 42
12 45
13 80
14 21
15 210

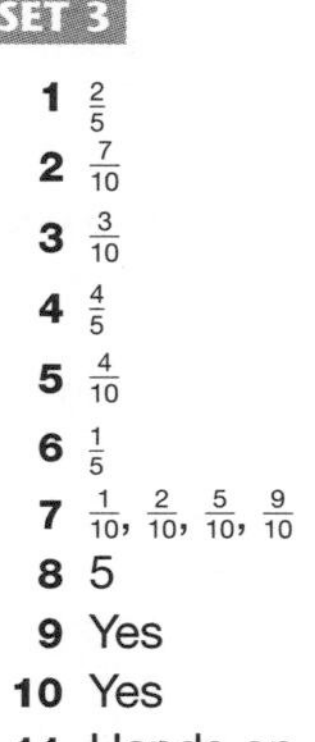

SET 3

1 $\frac{2}{5}$
2 $\frac{7}{10}$
3 $\frac{3}{10}$
4 $\frac{4}{5}$
5 $\frac{4}{10}$
6 $\frac{1}{5}$
7 $\frac{1}{10}, \frac{2}{10}, \frac{5}{10}, \frac{9}{10}$
8 5
9 Yes
10 Yes
11 Hands on

SET 4

1 C
2 10c each
3 24
4 16
5 $17.86
6 600
7 726
8 22 days
9 $15
10 6
11 $10.50
12 987
13 7
14 48
15 Seven thousand, two hundred and sixty
16 2

Space

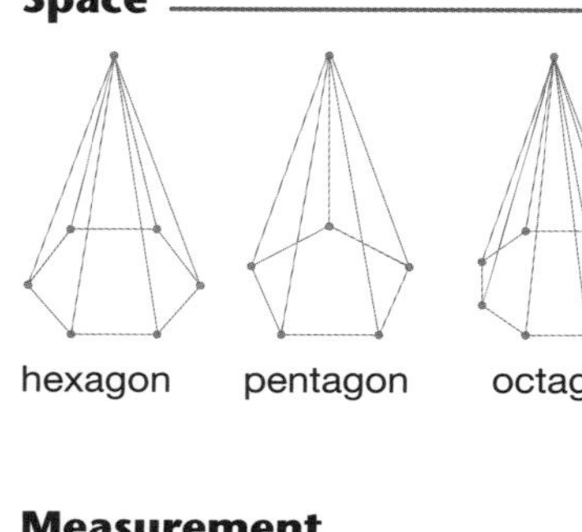

hexagon pentagon octagon

Measurement

1 14 cm^2
2 Hands on

UNIT 9 Number and Algebra

SET 1

1 21
2 20
3 2
4 18
5 20
6 33
7 32
8 20
9 July
10 48
11 4 each
12 25
13 3
14 1:25 pm
15 $8

SET 2

1 60 + 11 = 71
2 40 + 15 = 55
3 70 + 7 = 77
4 90 + 9 = 99
5 90 + 12 = 102
6 100 + 8 = 108
7 150 + 11 = 161
8 190 + 11 = 201
9 True
10 False (answer is 69)
11 True
12 True

SET 3

1 28
2 36
3 44
4 84
5 160
6 120
7 70
8 110
9 130
10 90
11 230
12 120
13 72
14 55
15 192
16 96

SET 4

1 $9.90
2 48
3 270
4 10
5 46
6 $22.95
7 80
8 148
9 24
10 1390
11 16
12 50c each
13 $4
14 72
15 One thousand, nine hundred and sixty-four
16 32

Statistics and Probability

1 Walk
2 Bike
3 5
4 Bike and train

Measurement

1 5:00 am
2 10:15 am
3 12:25 am
4 11:40 pm
5 5:45 pm
6 1:30 pm

Answers

UNIT 10 Number and Algebra

SET 1

1 13 cm
2 1
3 32
4 20c
5 30
6 12
7 14
8 32
9 October
10 22
11 28
12 18
13 32
14 9
15 7137
16 9

SET 2

1 442
2 212
3 621
4 231
5 102
203

SET 3

1 2

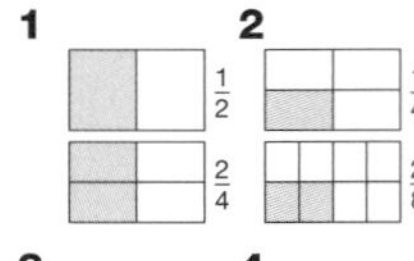

3 4

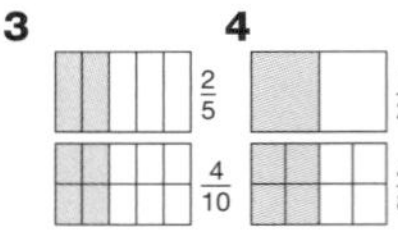

5 True
6 False
7 False
8 True
9 True
10 False

SET 4

1 61
2 $84
3 About 940
4 965
5 75
6 320
7 25
8 79
9 8752
10 44
11 427
12

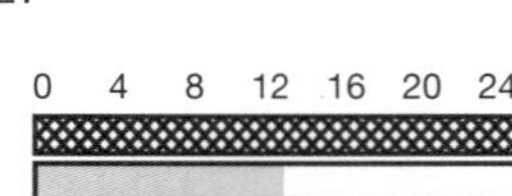

Space

Hands on

Measurement

1 70 mm
2 80 mm
3 50 mm
4 90 mm
5 55 mm

UNIT 11 Number and Algebra

SET 1

1 16
2 58
3 8
4 11
5 12
6 0
7 40
8 42
9 6
10 45
11 14
12 4
13 15
14 4 o'clock
15 963

SET 2

1 6, 7, 8, 9, 10, 11, 12, 13
2 10, 11, 12, 13, 14, 15, 16, 17
3 4, 8, 12, 16, 20, 24, 28, 32
4

hours	1	2	3	4	5	6	7
km	4	8	12	16	20	24	28

Answer: 28 km

5

boxes	1	2	3	4	5	6	7
cans	8	16	24	32	40	48	56

Answer: 56 cans

SET 3

1 1/3 2 2/3

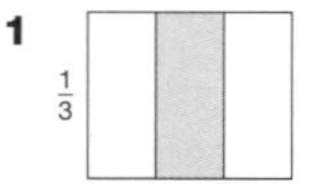

3 2/6 4 4/6

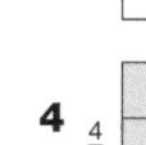
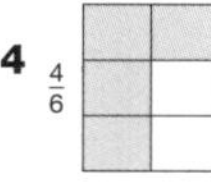

5 1/3

6 2/3

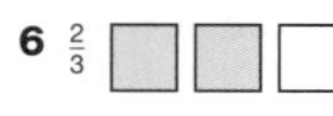

7 1/6
8 5/6

9 False
10 True
11 False
12 True

SET 4

1 368
2 956
3 7631
4 $25 each
5 28
6 7000
7 3000
8 7026
9 243
10 2
11 270
12 $56
13 $36
14 $20

Statistics and Probability

1

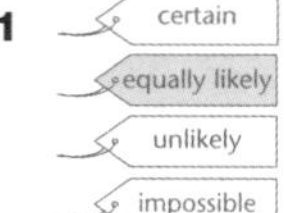

2

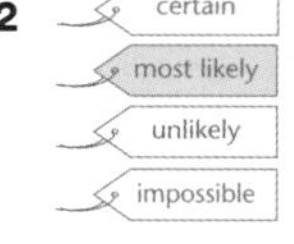

3 Blue and green

Measurement

1 A bucket full of water
2 A TV set
3 1 kg mass
4 2 kg bag of potatoes
5 1 kg mass
6 1 kg bag of feathers
7 1 kg mass

UNIT 12 Number and Algebra

SET 1

1 15
2 24
3 14
4 29
5 21
6 12
7 70
8 36c
9 19 kg
10 22
11 15
12 30
13 19
14 4
15 639
16 $8

SET 2

1	64	66	68	70	72	74	76	78
2	34	38	42	46	50	54	58	62
3	25	30	35	40	45	50	55	60
4	130	120	110	100	90	80	70	60
5	250	270	290	310	330	350	370	390
6	46	52	58	64	70	76	82	88
7	450	470	490	510	530	550	570	590
8	14	15	16	17	18	19	20	21
9	4	8	12	16	20	24	28	32
10	1	3	5	7	9	11	13	15

SET 3

1 24
2 40
3 56
4 72
5 80
6 48
7 32
8 88
9 16
10 96
11 12
13 14

SET 4

1 18
2 300
3 45
4 About 1460
5 $6.64
6 420, 435
7 33rd
8 18
9 $17 each
10 $200
11 400
12 $21.86
13 379
14 580
15 42
16 93
17 12

Space

Hands on

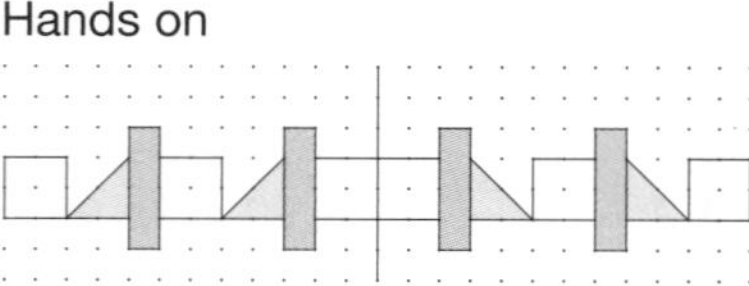

Statistics and Probability

Hands on

UNIT 13 Number and Algebra

SET 1

1 9
2 12
3 28
4 30
5 32
6 48
7 24
8 21c
9 31
10 386
11 7
12 5
13 25
14 1 thousand + 3 hundreds + 7 tens + 6 ones
15 926

SET 2

1 53
2 53
3 55
4 37
5 49
6 45
7 43
8 45
9 114
10 $47
11 $41

SET 3

Possible solutions:

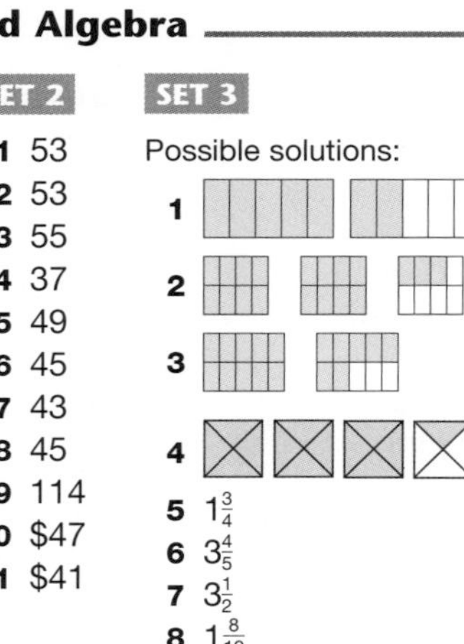

5 $1\frac{3}{4}$
6 $3\frac{4}{5}$
7 $3\frac{1}{2}$
8 $1\frac{8}{10}$

SET 4

1 35c each
2 40
3 $2.20
4 900
5 About 1800
6 1590
7 1600
8 3000
9 12
10 29
11 9, 18, 27, 36, 45, 54, 63
12 $1.35
13 6699
14 $240

Space

Hands on. Some examples below:

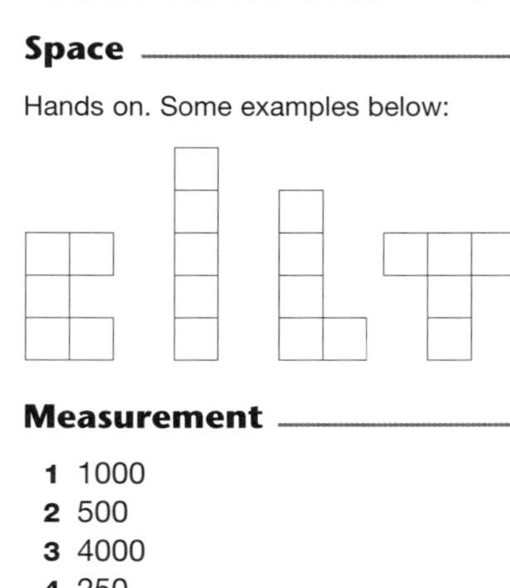

Measurement

1 1000
2 500
3 4000
4 250
5 3500
6 5000
7 250 mL
8 500

UNIT 14 Number and Algebra

SET 1

1 22
2 20
3 12
4 32 cm
5 21
6 15
7 21
8 50c
9 25 kg
10 15
11 21
12 24
13 11
14 28
15 $11.00

SET 2

1 $1\frac{2}{4}$, $1\frac{3}{4}$
2 3, $3\frac{1}{2}$
3 1, $1\frac{1}{5}$
4 $1\frac{1}{8}$, $1\frac{2}{8}$
5 $\frac{1}{10}$, $\frac{3}{10}$, $\frac{7}{10}$
6 $\frac{1}{4}$, $\frac{2}{4}$, $\frac{3}{4}$
7 $\frac{3}{8}$, $\frac{5}{8}$, $\frac{7}{8}$
8 $1\frac{1}{4}$, $1\frac{2}{4}$, $1\frac{3}{4}$
9 $2\frac{1}{5}$, $2\frac{2}{5}$, $2\frac{3}{5}$
10 True
11 False
12 True

SET 3

1 20
2 9
3 18
4 32
5 6
6 10
7 8
8 5 each
9 4
10 5
11 10
12 20
13 4

SET 4

1 E
2 5
3 20c each
4 67c
5 61
6 6000
7 About 1500
8 $4.50
9 4000
10 No
11 20
12 $170
13 180
14 35
15 Three thousand and sixty-two
16 45

Space

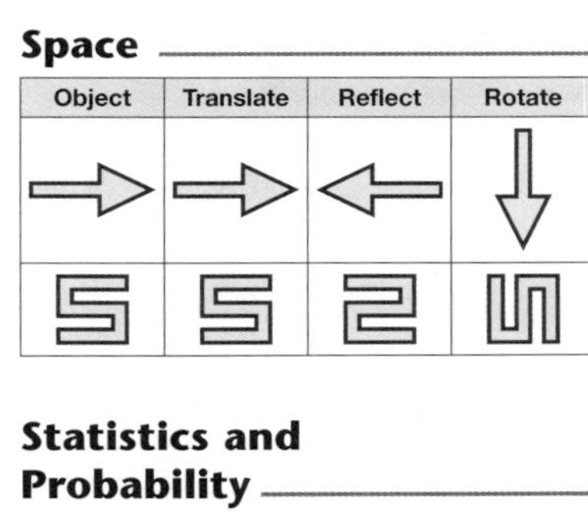

Statistics and Probability

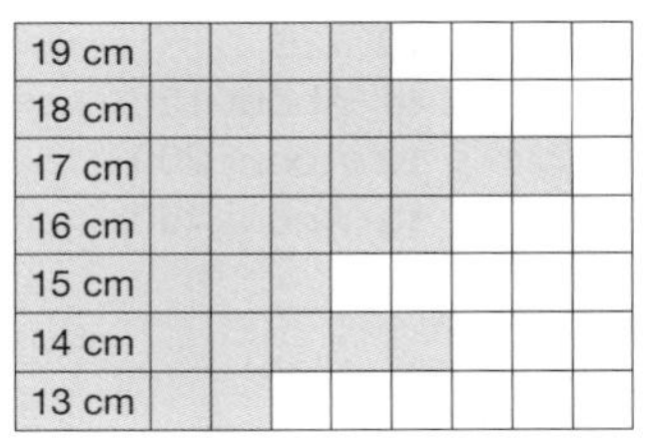

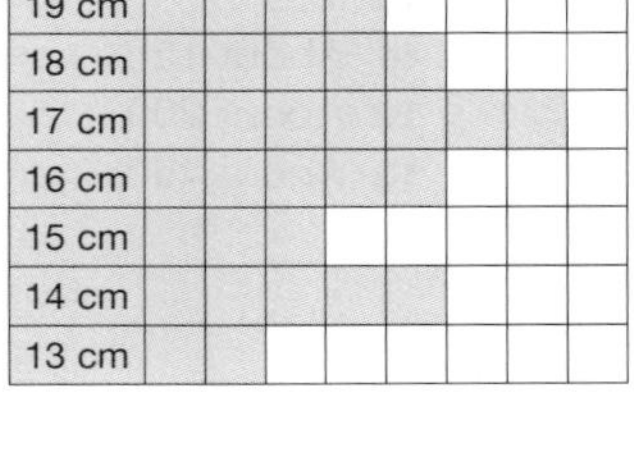

UNIT 15 Number and Algebra

SET 1

1 22
2 42
3 14
4 $24
5 9
6 16
7 40c
8 88
9 55 kg
10 44
11 18
12 10
13 15
14 3:55 pm
15 6 hundreds + 9 tens + 5 ones
16 15c

SET 2

1 $\frac{4}{10}$, 0.4
2 $\frac{6}{10}$, 0.6
3 $\frac{3}{10}$, 0.3
4 $\frac{9}{10}$, 0.9
5

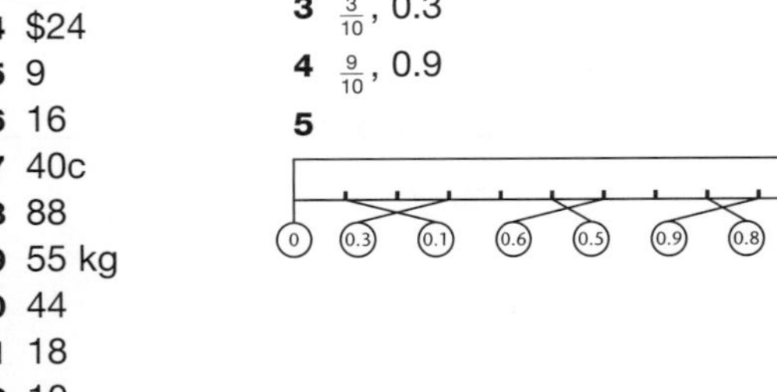

SET 3

1 hundreds
2 tens
3 ones
4 thousands
5 hundreds
6 tens of thousands
7 thousands
8 23 296
9 35 413
10 54 086

SET 4

1 $23.80
2 $4.60
3 $9.75
4 54th
5 42
6 54, 72, 90, 108
7 $1.32
8 $1.55
9 1, 2, 3, 4, 6, 9, 12, 18, 36
10 350
11 Green Point, 26
12 580
13 9

Space

Hands on

Measurement

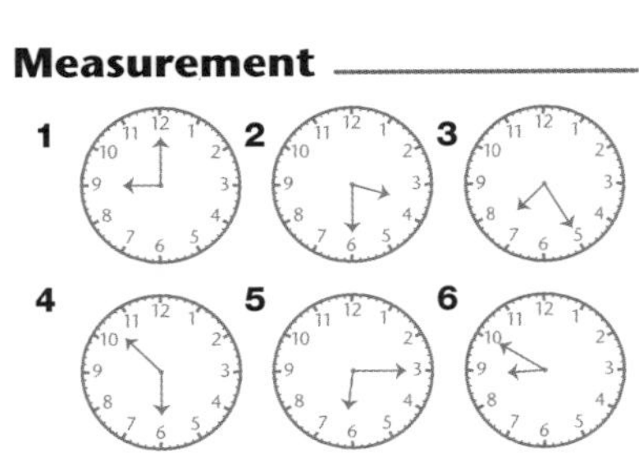

Answers

UNIT 16 Number and Algebra

SET 1

1 42
2 37
3 33
4 43
5 28
6 18
7 35
8 48
9 22
10 10
11 14
12 4
13 63
14 18
15 $2.60

SET 2

1 564
2 773
3 774
4 861
5 890
6 752
7 193
8 375
9 167
10 579
11 $422

SET 3

	Numbers	× 9	× 3	× 6
1	3	27	9	18
2	7	63	21	42
3	10	90	30	60
4	2	18	6	12
5	6	54	18	36
6	5	45	15	30
7	8	72	24	48

8 $72
9 81
10 90

SET 4

1 9620
2 25
3 210
4 $7 each
5 358, 367
6 $40
7 $5.75
8 436th
9 16
10 1, 2, 3, 4, 6, 8, 12, 24
11 $48
12 52
13 330
14 $8 \times 3 \times 2 = 48$
or
$8 \times 2 \times 3 = 48$

Statistics and Probability

Fruit — strawberry, bananas, pear
Toys — teddy bear, doll, marbles
Jewellery — ring, necklace

Measurement

1 50 g
2 25 g
3 35 g
4 75 g
5 125 g

UNIT 17 Number and Algebra

SET 1

1 47
2 89
3 22
4 41c
5 42
6 21
7 28
8 35
9 19
10 12
11 20
12 893
13 3
14 64
15 $39

SET 2

1 215
2 108
3 228
4 125
5 124
6 112
7 180
8 168
9 222
10 126
11 About 120
12 About 200
13 About 490

SET 3

1 $5 \times 5 = 25$
2 $32 \div 8 = 4$ and $8 \times 4 = 32$
3 $40 \div 10 = 4$ and $10 \times 4 = 40$
4 $28 \div 7 = 4$ and $7 \times 4 = 28$
5 $81 \div 9 = 9$ and $9 \times 9 = 81$
6 $42 \div 7 = 6$ and $7 \times 6 = 42$
7 $36 \div 6 = 6$ and $6 \times 6 = 36$
8 $54 \div 6 = 9$ and $6 \times 9 = 54$
9 7 remainder 3
10 3 remainder 2

SET 4

1 36
2 257
3 About 550
4 10
5 $14.95
6 $3.10
7 1824
8 $33.50
9 350
10 $4.30
11 1327, 1427
12 $14
13 44
14 $129
15 Two thousand, four hundred and ninety-five
16 9750
17 No

Space

1 North
2 West
3 South-west
4 North-west
5 North
6 North-east

Measurement

1 16 cm
2 12 cm
3 14 cm
4 18 cm

UNIT 18 Number and Algebra

SET 1

1 14
2 35
3 19
4 $14
5 16
6 14
7 70c
8 89
9 63 kg
10 72
11 29
12 20
13 14
14 9 hundreds + 6 tens + 2 ones
15 3:22

SET 2

1 4567
2 2835
3 $3952
4 6050
5 3156
6 True
7 True
8 False

SET 3

1 8, 16, 24, 32, 40, 48, 56, 64
2 16, 17, 18, 19, 20, 21, 22, 23
3 4, 8, 12, 16, 20, 24, 28, 32
4 18, 24, 30, 36, 42, 48, 54, 60
5 Rule: add 4
5, 6, 7, 8, 9, 10, 11, 12
6 Rule: multiply by 3
3, 6, 9, 12, 15, 18, 21, 24

SET 4

1 237
2 4000
3 6
4 14
5 4996
6 39
7 $39
8 30
9 1400
10 5
11 42
12 4

Space

Top Front Side

Measurement

Hands on

UNIT 19 Number and Algebra

SET 1

1 16
2 13
3 42
4 28
5 56
6 49
7 62
8 46
9 20
10 18
11 586
12 13
13 56
14 7
15 50c

SET 2

	Nearest 100	Nearest 1000
1	1000	1000
2	800	1000
3	1200	1000
4	1800	2000
5	2400	2000
6	7600	8000
7	4200	4000

8 300 + 700 = 1000
9 200 + 800 = 1000
10 400 + 900 = 1300
11 900 + 1000 = 1900
12 900 + 800 = 1700
13 1000 + 1000 = 2000

SET 3

Questions 1–6: any 4 of the factors given

1 1, 2, 4, 5, 10, 20
2 1, 2, 4, 8, 16
3 1, 2, 3, 4, 6, 12
4 1, 2, 4, 5, 8, 10, 20, 40
5 1, 2, 4, 8, 16, 32
6 1, 2, 3, 5, 6, 10, 15, 30
7 Yes
8 Yes
9 No
10 Yes
11 1, 2, 3, 4, 6, 8, 12, 24

SET 4

1 483
2 4 m 37 cm
3 18
4 3500
5 2
6 7
7 $6 each
8 8546
9 42
10 54
11 About 390
12 3999
13 $5
14 Length = 40 cm
Width = 20 cm

Space

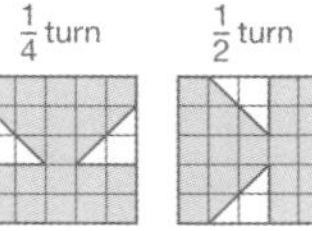

Statistics and Probability

1 A
2 C
3 C
4 B
5 A

UNIT 20 Number and Algebra

SET 1

1 16
2 18
3 10
4 29
5 21c
6 21
7 927
8 3
9 29
10 420
11 $7 each
12 $88
13 56 kg
14 Two thousand and sixty

SET 2

1 116
2 571
3 365
4 178
5 $305
6 $788
7 702
8 280
9 78
10 165
11 62

SET 3

1 22
2 43
3 47
4 12
5 31
6 14
7 9
8 8
9 5
10 7
11 13
12 15
13 3
14 4

c

	7		15	
4		12		47

SET 4

1 32
2 1258
3 8
4 1706
5 $24.20
6 $3.05
7 50
8 $10.95
9 140
10 $1.17
11 $7.50
12 $77
13 9950
14 7
15 28 cm
16 7

Space

1 b
2 b

Measurement

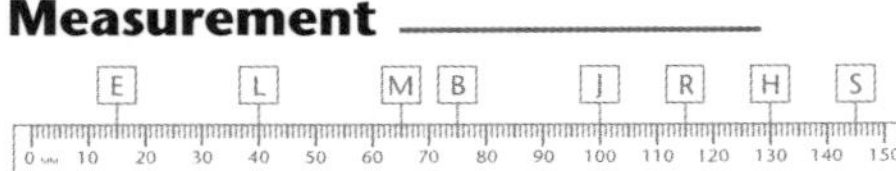

UNIT 21 Number and Algebra

SET 1

1 10
2 8
3 18
4 36
5 14
6 48
7 36
8 21
9 21
10 17
11 $3.70
12 6
13 25
14 5
15 4 thousands + 0 hundreds + 5 tens + 7 ones

SET 2

	Becomes	Subtract	Answer
1	23 + 40 = 63	2	61
2	43 + 50 = 93	1	92
3	34 + 50 = 84	2	82
4	27 + 60 = 87	1	86
5	32 + 70 = 102	2	100
6	117 + 40 = 157	1	156

	Becomes	Add	Answer
7	27 + 40 = 67	1	68
8	44 + 30 = 74	2	76
9	67 + 20 = 87	1	88
10	48 + 50 = 98	2	100
11	56 + 30 = 86	3	89
12	137 + 40 = 177	2	179

SET 3

1 Any 25 squares
2 Any 63 squares
3 Any 55 squares
4 Any 22 squares
5 $\frac{33}{100}$
6 $\frac{47}{100}$
7 $\frac{69}{100}$
8 $\frac{98}{100}$

SET 4

1 3000
2 337
3 405
4 36
5 100
6 12
7 128
8 1839
9 $16
10 92
11 $5.50
12 $3.25
13 7 r 2
14 100 cm and 80 cm

Space

Hands on. Some examples below:

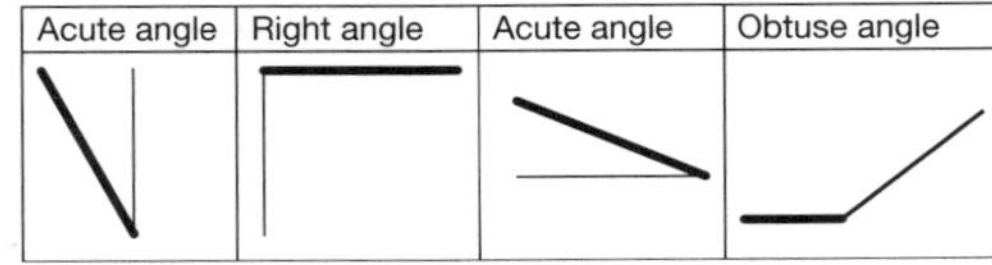

Measurement

1 225 mL
2 50 mL
3 500 mL
4 750 mL
5 275 mL
6 700 mL

Answers

UNIT 22 Number and Algebra

SET 1

1 35
2 15
3 3
4 35
5 8
6 10
7 30
8 24
9 3
10 105
11 3
12 407
13 14
14 3 thousands + 2 hundreds + 0 tens + 4 ones
15 Two thousand and five

SET 2

1 $505
2 $5
3 $39
4 380
5 260 g
6 55 marbles
7 880
8 141
9 236
10 616
11 181
12 479

SET 3

1 6
2 3
3 5
4 4
5 4
6 3
7 24
8 $18

SET 4

1 10
2 19
3 $33
4 $2471
5 8 each
6 64
7 4
8 4837
9 50
10 432nd
11 $8.35
12 143
13 15
14 48
15 Four thousand, three hundred and nine
16 $3.60

Space

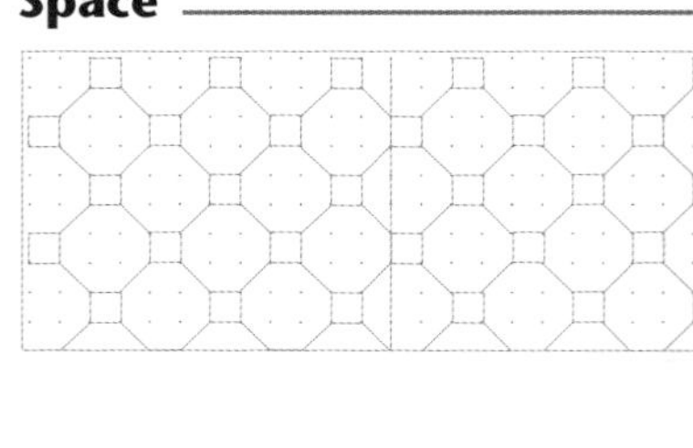

Measurement

Hands on

UNIT 23 Number and Algebra

SET 1

1 48
2 9
3 25
4 10
5 40
6 19
7 34
8 25
9 80
10 3, 6, 9, 12, 15, 18, 21
11 5
12 40
13 8
14 202
15 6 thousands + 5 hundreds + 2 tens + 0 ones
16 $35.05

SET 2

1 92
2 120
3 105
4 138
5 96
6 144
7 70
8 114
9 84
10 $175
11 $185

SET 3

1 $\frac{31}{100}$
2 $\frac{68}{100}$
3 $\frac{13}{100}$
4 $\frac{80}{100}$
5 0.15
6 0.21
7 0.25
8 0.73
9 0.99

SET 4

1 3187
2 4000
3 2 m 47 cm
4 4
5 35 cm
6 6
7 $6
8 $4 each
9 14
10 $3.15
11 $1.80
12 5:30
13 Yes
14 35
15 Four hundred and ninety-seven
16 25 min
17 3700

Space

At the skull

Statistics and Probability

1 $6
2 Alex
3 $2
4 $8
5 $4
6 $23

UNIT 24 Number and Algebra

SET 1

1 24
2 8
3 38
4 5
5 15
6 32
7 62
8 36
9 32
10 70, 60, 50
11 6
12 38
13 60 kg
14 3 thousands + 0 hundreds + 7 tens + 6 ones
15 7, 16

SET 2

	Becomes	Subtract	Answer
1	26 + 50 = 76	1	75
2	37 + 60 = 97	2	95
3	41 + 90 = 131	1	130
4	43 + 50 = 93	3	90
5	13 + 70 = 83	1	82
6	114 + 50 = 164	3	161

	Becomes	Add	Answer
7	33 + 70 = 103	1	104
8	25 + 40 = 65	2	67
9	61 + 30 = 91	3	94
10	76 + 20 = 96	2	98
11	66 + 60 = 126	3	129
12	123 + 50 = 173	1	174

SET 3

1 4
2 7
3 5
4 3
5 6
6 9
7 2
8 8
9 10
10 8
11 4
12 1
13 5
14 11
15 3
16 10
17 1
Rudolf has a red nose
18 8

SET 4

1 4
2 142
3 70
4 19
5 8
6 Yes
7 $5.95
8 1, 2, 3, 6, 9, 18
9 $24
10 1 L 375 mL
11 Yes
12 0 tens
13 5 coins ($2, $2, 50c, 10c, 5c)
14 200 km
15 400 km

Statistics and Probability

1
cakes
biscuits
pancakes
popcorn

2 Cakes
3 31

Measurement

1 1 kg 600 g
2 2 kg 260 g
3 4 kg 950 g
4 5 kg 20 g
5 3 kg 35 g

UNIT 25 Number and Algebra

SET 1

1 49
2 11
3 23
4 7
5 56
6 17
7 63
8 70
9 29
10 28
11 13
12 4
13 259
14 0 thousands
+ 3 hundreds
+ 2 tens
+ 0 ones
15 44

SET 2

1 900
2 388
3 495
4 778
5 995
6 933
7 881
8 1065
9 $759

SET 3

1 $\frac{80}{100}$
2 1.0
3 0.21
4 0.43
5 $\frac{33}{100}$
6 4.01
7 1 hundredth
8 1 whole
9 2 tenths
10 7 hundredths
11 0.09, 0.29, 0.35
12 0.29, 0.33, 0.39
13 0.87, 1.07, 7.01
14 8.59, 9.58, 9.85

SET 4

1 About 340
2 3250
3 8
4 500 m
5 6 r 4
6 7
7 30
8 28
9 44
10 4885
11 143
12 54
13 9654
14 24

Space

1

2

Measurement

	Less than 1 m^2	About 1 m^2	More than 1 m^2
TV screen	✓		
Notice board		✓	
Skateboard		✓	
Door			✓
Photo frame	✓		
Bedroom floor			✓

UNIT 26 Number and Algebra

SET 1

1 14
2 15
3 47
4 10
5 64
6 48
7 70
8 50
9 24
10 8
11 128
12 20
13 379
14 One hundred and seventy-six
15 24

SET 2

1 0.53
2 0.85
3 0.21
4 0.32
5 0.54
6 0.68
7 0.64
8 0.75
9 0.96
10 0.72, 0.83, 0.85, 0.91

SET 3

1 23
2 16
3 31
4 81
5 45
6 69
7 17

SET 4

1 $534
2 4250
3 1, 3, 5, 15
4 100
5 8 r 2
6 10
7 45
8 2000
9 39
10 1, 2, 4, 5, 10, 20
11 $1.80
12 36
13 1267
14 $3.80
15 37

Space

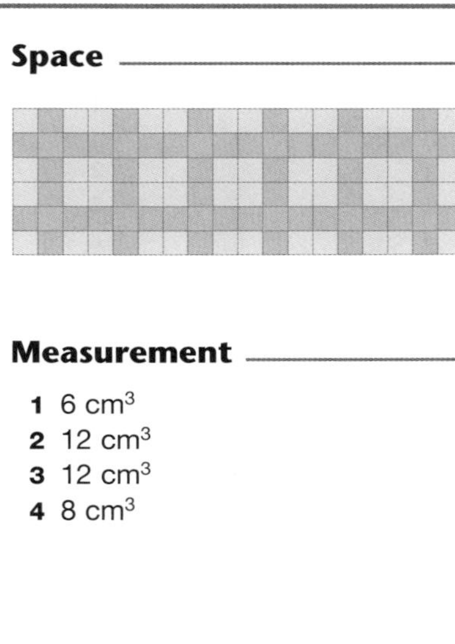

Measurement

1 6 cm^3
2 12 cm^3
3 12 cm^3
4 8 cm^3

UNIT 27 Number and Algebra

SET 1

1 25
2 2
3 7
4 33
5 24
6 27
7 45
8 36
9 63
10 30
11 10
12 $3.09
13 15
14 0 thousands
+ 0 hundreds
+ 2 tens + 7 ones
15 9

SET 2

	×	0	5	3	6	4	7	9
1	3	0	15	9	18	12	21	27
2	5	0	25	15	30	20	35	45
3	7	0	35	21	42	28	49	63
4	9	0	45	27	54	36	63	81
5	8	0	40	24	48	32	56	72

6 336
7 72
8 300
9 210
10 $228

SET 3

1 9
2 12
3 16
4 21 each
5 $9 each
6 $12
7 $14
8 $30
9 $26
10 $32
11 14
12 18
13 13 r 4
14 13 r 4

SET 4

1 0
2 8
3 30
4 3
5 Yes
6 0.21
7 9000
8 $1.05
9 $3
10 1, 2, 3, 5, 6, 10, 15, 30
11 5500
12 $60
13 36 km
14 56 ÷ 4 ÷ 2 = 7
or
56 ÷ 2 ÷ 4 = 7

Space

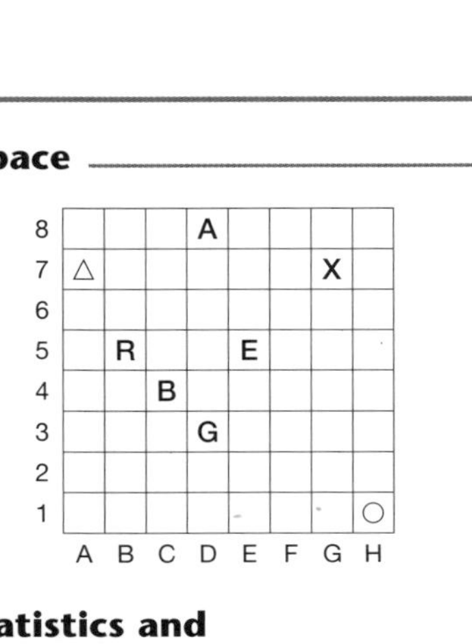

Statistics and Probability

1 Red
2 Blue and pink
3 Green

Answers

UNIT 28 Number and Algebra

SET 1

1 14
2 24
3 5
4 6, 12, 18, 24, 30
5 244
6 9
7 19
8 5
9 365
10 7
11 7
12 27
13 33
14 $6.59
15 4321
16 $48

SET 2

1 8781
2 7984
3 2995
4 8974
5 9572
6 6894
7 4125
+3863
7988
8 6475
+3320
9795

SET 3

1 10, 10
2 17, 17
3 90, 90
4 150, 150
5 132, 132
6 276, 276
7 No
8 30, 30
9 28, 28
10 24, 24
11 48, 48
12 35, 35
13 32, 32
14 No

SET 4

1 $10.75
2 $25.20
3 56
4 1940
5 17
6 1, 4, 9, 16, 25, 36, 49
7 $2.55
8 85
9 9 r 3
10 6000
11 $297
12 80
13 About 440
14 5
15 $12

Statistics and Probability

1
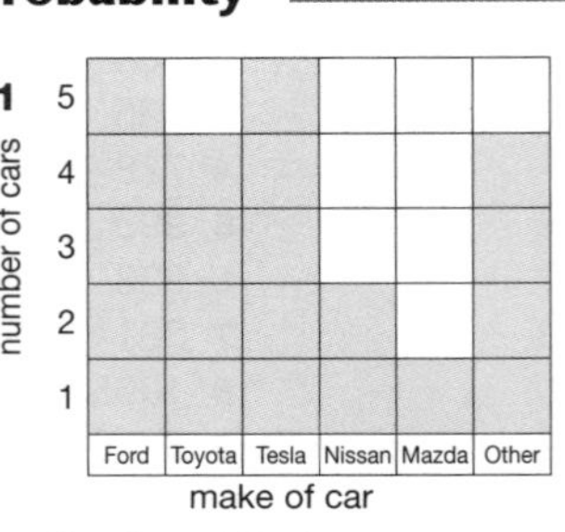

2 Ford and Tesla

Measurement

1 1 kg 400 g
2 2 kg 800 g
3 1 kg 800 g
4 2 kg 300 g

UNIT 29 Number and Algebra

SET 1

1 40c
2 20c
3 $6
4 $10
5 $30
6 $2.50
7 $30
8 $7
9 66
10 29
11 7
12 $9
13 $8 each
14 Nine dollars and fifty-six cents
15 1 pizza or 2 half pizzas

SET 2

1 16, 25, 36, 49, 64
2 45
3 27
4 48
5 25
6 40
7 7
8 140
9 225
10 54
× 6
324

SET 3

1 3
2 5 r 4
3 11
4 13
5 8
6 6
7 8 r 1
8 9
9 9 r 2
10 10 r 1
Correct bingo card is **A**
11 22 r 2
12 15 r 2
13 13 r 1
14 38 r 1
15 16 r 1
16 17 r 1

SET 4

1 0
2 0.37
3 350
4 $2.70
5 5
6 1, 3, 7, 21
7 6
8 $12.90
9 Yes
10 Half
11 70
12 0.3
13 10
14 125
15 104 cm
16

22	8	18
12	16	20
14	24	10

17 $35.00

Number and Algebra

1 $1.85
2 $2.45
3 $1.80
4 $0.90
5 $1.15

Statistics and Probability

Facebook	Instagram	X	TikTok
48	24	12	12

UNIT 30 Number and Algebra

SET 1

1 55c
2 17c
3 $6
4 $3
5 $91
6 $1.50
7 $30
8 $3
9 24
10 50
11 51
12 $2
13 $8 each
14 Sixteen dollars and fifty-eight cents

SET 2

1 8092
2 7274
3 9355
4 5680
5 9254
6 8294
7 5000
+5000
10000

SET 3

	Tenths	Hundredths	Decimal
1	$\frac{1}{10}$	$\frac{10}{100}$	0.10
2	$\frac{3}{10}$	$\frac{30}{100}$	0.30
3	$\frac{9}{10}$	$\frac{90}{100}$	0.90
4	$\frac{5}{10}$	$\frac{50}{100}$	0.50
5	$\frac{7}{10}$	$\frac{70}{100}$	0.70
6	$\frac{2}{10}$	$\frac{20}{100}$	0.20
7	$\frac{6}{10}$	$\frac{60}{100}$	0.60

8 True
9 True
10 True
11 False
12 True
13 True
14 False
15 False
16 True

SET 4

1 $79
2 4000
3 1, 2, 3, 4, 6, 9, 12, 18, 36
4 250 mL
5 5c each
6 100
7 6:35 am
8 31
9 50
10 1, 2, 3, 4, 6, 8, 12, 24
11 $4.20
12 48
13 Sally

Number

1 $2965
2 2201
3 7700
4 $4638

Statistics and Probability

1 $160
2 $120
3 $20
4 10 June
5 20 June
6 No

UNIT 31 Number and Algebra

SET 1

1 11
2 48
3 14
4 56
5 32
6 10
7 30
8 81
9 8
10 $13
11 1350
12 2 thousands + 7 hundreds + 0 tens + 6 ones
13 22
14 30
15 36

SET 2

1 32
2 Yes
3 Yes
4 42
5 70
6 92
7 156
8 125
9 105
10 325
11 114
12 $240

SET 3

1	0.5	0.7	0.2
2	0.6	0.9	0.3
3	0.34	0.58	0.41
4	2.30	2.75	2.12
5	5.95	4.62	5.34
6	7.62	6.84	8.11

7 2 m
8 1 m
9 1 m
10 5 m
11 No

SET 4

1 1250
2 14 each
3 $205
4 10
5 False
6 1378
7 20
8 0.37
9 1.3
10 325
11 No
12 $66
13 $15
14 $55

Space

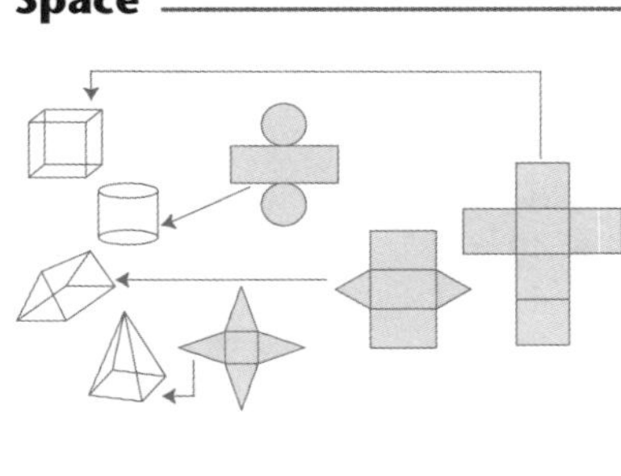

Measurement

4th	2nd	1st	3rd	5th
Eboni	Caleb	Henri	Jack	Jess

UNIT 32 Number and Algebra

SET 1

1 30
2 24
3 13 mL
4 30c
5 $1.50
6 2 m 28 cm
7 7 kg
8 64
9 7 m
10 81
11 50c
12 7 m
13 45 mL
14 84
15 16

SET 2

1

Multiply by 5							
2	4	6	8	10	12	14	16
10	20	30	40	50	60	70	80

2

Adding 17							
1	2	3	4	5	6	7	8
18	19	20	21	22	23	24	25

3

Multiply by 4 then add 1							
1	2	3	4	5	6	7	8
5	9	13	17	21	25	29	33

4

Multiply by 6 then subtract 2							
3	4	5	6	7	8	9	10
16	22	28	34	40	46	52	58

5

Multiply by 4 then subtract 3							
1	2	3	4	5	6	7	8
1	5	9	13	17	21	25	29

6

Multiply by 7 then add 4							
1	2	3	4	5	6	7	8
11	18	25	32	39	46	53	60

SET 3

1 ✓
2 ✗
3 ✓
4 ✗
5 ✓
6 ✓
7 $\frac{3}{4}$, 1, $1\frac{1}{4}$, $1\frac{2}{4}$
8 1.6, 1.8, 2.0, 2.2

SET 4

1 21
2 True
3 5000
4 427
5 800
6 0.7
7 0.17
8 $4.50
9 No
10 450 km
11 1, 5, 7, 35
12 0.07
13 10, 12, 6, 12, 22, 20, 12, 4, 36

Statistics and Probability

1 OB and MT
2 GG
3 MD
4 100

Measurement

1 9 m^2
2 50 m^2

UNIT 33 Number and Algebra

SET 1

1 18
2 20
3 24
4 11
5 39
6 22
7 27
8 21
9 35
10 31
11 45
12 $24
13 17
14 8 thousands + 7 hundreds + 5 tens + 6 ones
15 $1.80

SET 2

	Numbers	× 10	× 100	× 1000
1	5	50	500	5000
2	8	80	800	8000
3	19	190	1900	19 000
4	27	270	2700	27 000
5	45	450	4500	45 000
6	61	610	6100	61 000
7	77	770	7700	77 000
8	80	800	8000	80 000
9	1	10	100	1000
10	99	990	9900	99 000

11 48
12 5
13 21
14 <
15 >

SET 3

Hands on. Possible solutions:

1 7 + 3 + 9 = 19
2 6 + 4 + 8 = 18
3 18 + 2 + 5 = 25
4 9 + 6 + 5 = 20
5 5 × 2 × 3 = 30
6 5 × 2 × 6 = 60
7 5 × 4 × 3 = 60
8 23
9 60
10 50
11 69
12 72
13 14

SET 4

1 14
2 50
3 16 r 2
4 7
5 3
6 7010
7 1, 2, 4, 8, 16, 32
8 5542
9 12
10 Yes
11 $24.50
12 6300
13 About 500
14 $735

Statistics and Probability

Favourite sports

24, 21, 18, 15, 12, 9, 6, 3
N F G S A

Measurement

1 1.2 L
2 3.5 L
3 5.0 L

Answers

UNIT 34 Number and Algebra

SET 1

1 15
2 43
3 25
4 8
5 40
6 4
7 24
8 2
9 24
10 $21
11 1603c
12 36
13

SET 2

1 2
2 4
3 2
4 3
5 2
6 4
7 4

SET 3

1 56
2 45
3 27
4 48
5 25
6 40
7 7
8 140
9 225
10 504
11 588
12

HUND	TENS	ONES
	$^{2}5$	4
		6
3	2	4

SET 4

1 298
2 62
3 171
4 3316
5 6
6 0.8
7 $4.05
8 Four thousand, one hundred and seven
9 144
10 No
11 $3.60
12 1, 2, 4, 5, 10, 20
13 $952
14 4164
15 11
16 5

Space

1 3
2 yes
3 Yes
4 2
5 Tulip St/ Fern St or Waratah Rd/ Daffodil Dr

Measurement

1 200 mL
2 100 mL
3 350 mL
4 450 mL

UNIT 35 Number and Algebra

SET 1

1 $42
2 $4 each
3 11
4 $32
5 52
6 51
7 $20
8 1465c
9 56
10 27
11 39
12 6 kg
13 18
14 6
15 90, 80, 70, 60, 50, 40, 30
16 20

SET 2

1 4314
2 4219
3 4323
4 2456
5 5364
6 5189
7 500
8 750
9 $4300
10 $3225

SET 3

1 No
2 No
3 Yes
4 No
5 No
6 No
7 Yes

SET 4

1 37
2 $45.00
3 375
4 Yes
5 $9 each
6 63
7 $24.50
8 About 1400
9 20 mL
10 Length 9 m Width 8 m

Space

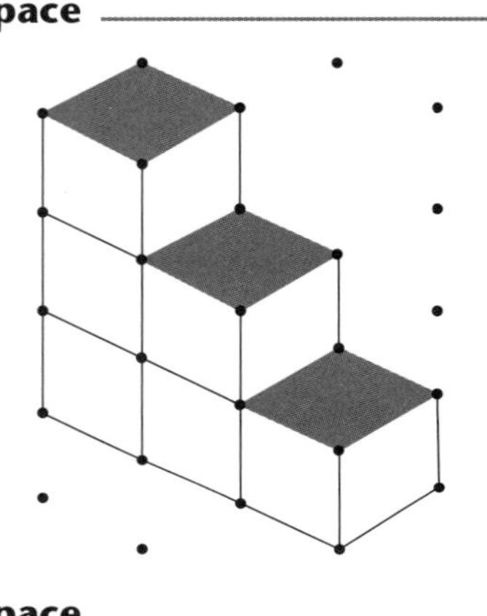

Space

1 (D,3)
2 (A,3)
3 (H,1)
4 (H,3)
5 (F,2)
6 (B,1)